The Future for
Ocean Technology

The Future for Science and Technology Series

As we move through the 1980s there is a whole galaxy of new technologies and new industries ready to emerge. They represent important opportunities and risks for business, so there is an acute need for publications which reflect the latest scientific thinking and which can be read and understood by decision-makers and their advisers in business and government. This series aims to meet the needs of entrepreneurs and managers searching for new business opportunities; public servants interested in developing schemes and regulations for science and industry; officials of trade unions and other voluntary organizations seeking to protect the interests of various groups in society.

The books are intended for an international market, and it is hoped that they will be influential in directing businessmen to explore and invest in new areas of technology; persuading technologists to expand their research into new fields; encouraging civil servants and politicians to support viable new initiatives; and helping various groups in society to assess the likely social consequences of new technological developments.

Series Editor: Bernard Taylor, Professor of Business Policy, Henley Management College, Henley-on-Thames, Oxon RG9 3AU.

Other titles in the Series

The Future for Automotive Technology by Ulrich Seiffert and Peter Walzer
The Future for Space Technology by Geoffrey K. C. Pardoe.

Also planned:

The Future of Telecommunications by Adrian Stokes, OBE and Neil Hollingham.

The Future for Ocean Technology

Glyn Ford, Chris Niblett
and Lindsay Walker

Frances Pinter (Publishers), London and
Wolfeboro, New Hampshire

First published in Great Britain in 1987 by
Frances Pinter (Publishers) Limited
25 Floral Street, London WC2E 9DS

First published in the United States of America in 1987 by
Frances Pinter (Publishers) Limited
27 South Main Street, Wolfeboro, NH 03814-2069

British Library Cataloguing in Publication Data
Niblett, Chris,
    The future for ocean technology.—
    (The Future for science and technology series)
    1. Ocean engineering
    I. Title   II. Walker, Lindsay   III. Ford,
    Glyn   IV. Series
    620´.4162   TC1645
    ISBN 0-86187-522-2

Library of Congress Cataloging in Publication Data
Niblett, C. A. (Chris A.), 1954–
    The future for ocean technology.
    (The Future for science and technology series)
    Bibliography: p.
    ·Includes index.
    1. Ocean engineering.   2. Marine mineral resources.
3. Ocean energy resources.   I. Walker, L. M.,
1945–   . II. Ford, J. G.   III. Title.   IV. Series.
TC1645.N53   1987   621.042´09162   86-25527
    ISBN 0-86187-522-2

Typeset by Joshua Associates Limited, Oxford
Printed by Biddles of Guildford Ltd

Contents

List of tables and figures   vi

Series Editor's Introduction   vii

Preface   xiii

Acknowledgements   xiv

1 Introduction   1

2 Manganese nodules   7

3 Phosphorites and muds   25

4 Other minerals from the sea   45

5 Ocean Thermal Energy Conversion   61

6 Energy from the sea: other proposals   79

7 Artificial islands   93

8 Surveying and sensing for marine technology   109

9 Summary and prospects   119

Bibliography   133

Index   137

List of tables and figures

Tables

2.1 Consortia involved in ocean mining: 1980   9
8.1 Aeroplane effectiveness as against two types of satellite   110

Figures

2.1 Manganese nodule mining and processing system   12
2.2 Continuous line bucket system   13
2.3 Types of nodule collector   16
3.1 World-wide distribution of sea-floor phosphorites   26
3.2 Phosphorite dredging by means of buckets   28
3.3 Suction pumping of muds   34
3.4 Red Sea Commission area   40
4.1 Sites of hydrothermal and polymetallic sulphide deposits 49
5.1 World-wide distribution of OTEC Thermal Resource   62
5.2 Schematic layout of OTEC systems   64
5.3 Possible platform configurations   68
5.4 Heat-exchanger types   71

Series Editor's Introduction

The sea covers 70 per cent of the earth's surface but, though it is such a vast resource, its potential has hardly been charted. The oceans represent a new frontier for science and mankind. This book examines the future prospects for ocean research and exploration, the technological possibilities, the business opportunities and the social and political implications.

OCEANOGRAPHY

Marine study has changed its emphasis and scope. It started with the study of winds, currents, sailing conditions and navigation, but the present emphasis is on oceanography. Marine studies include:

- communication among whales and dolphins;
- the use of plankton and krill as a source of protein;
- new approaches to fishing using acoustic devices such as sonars, which are particularly useful at deeper levels;
- the use of fishing submarines to trawl below the surface;
- marine fish farming of eels, mullet and prawns and fish ranching: transplanting young fish or shell fish in suitable areas;
- the use of desalination plants to provide fresh water where water is scarce;
- exploitation of minerals under the sea, especially oil, gas and coal but also the undersea mining of sulphur and the dredging of manganese nodules.

OCEAN TECHNOLOGY AND THE FUTURE

Marine technology is improving rapidly, bringing new mineral resources within reach and making food from the sea more readily available. Recent advances include the following:

- super-ships, such as super-tankers, have been developed;
- the deep-diving submarines allow men to descend to the ocean bottom;
- it is possible to drill for oil and gas in deep water with semi-submersible platforms and self-propelled vessels;
- ships can navigate very precisely using orbiting satellites, inertial guidance systems and electronic devices;

- it is possible to explore the ocean bottom using television and sonar;
- new lighter, stronger materials are being used for marine operations—steel and aluminium alloys, glass, nylon and plastics;
- new types of craft have appeared—such as the Hovercraft and the Sealab—and new techniques for diving.

ECONOMIC AND BUSINESS OPPORTUNITIES

The oceans' resources are too huge to comprehend: 150,000 species of animal, 16,000 types of fish, billions of tons of plankton, huge quantities of mineral salts, millions of tons of iron and manganese in the form of 'nodules', big reserves of gas and oil, also gold, nickel, silver, molybdenum and diamonds.

Although the average depth of the sea is about 4 kilometres, normal exploration is restricted to the continental shelf around 300 metres below the surface. New techniques of diving and 'aquanautics' are opening up the possibility of fish farming and mineral prospecting on these 'shelves'.

In business terms the first requirement is for aids to ocean communication and exploration: diving equipment and support systems, including submersibles, communication systems, navigational aids, cameras and television, power tools and also tractors, probes, tools for burying pipelines. Apart from diving systems and equipment for seabed exploration, a whole range of equipment is needed for use on the surface of the sea, on beaches and on ships.

However, even this is a narrow, short-term view. On a global basis the oceans offer a whole new arena for business—for the food industries, oil and gas, metal mining and manufacture, leisure, transportation and communication.

SOCIAL AND POLITICAL IMPLICATIONS

As the oceans are explored and exploited, so there is increasing competition between different nations and interest groups. This results in various social costs:

- the coasts are being polluted, built on and damaged by dumping, dredging and off-shore drilling;
- the seas in heavily populated enclosed areas such as the Mediterranean become polluted and remain polluted from human sewage, chemicals and rubbish;
- the wreckage of giant oil tankers is a constant menace to beaches and sea life;
- local fishing industries are declining because of over-fishing and the use of 'factory ships'.

As competition for the oceans' resources increases, it becomes more urgent to develop international agreements and controls, e.g. to set limits to national sovereignty 'off-shore', to enforce a code of practice for fishing and to control pollution, and the carriage of dangerous goods.

THE MARINE RESOURCES PROJECT

The Marine Resources group at Manchester University is one of Britain's leading research units concerned with ocean technology. In this book, Glyn Ford, Christopher Niblett and Lindsey Walker summarize their conclusions from ten years of research. They have focused on the mineral and energy resources of the oceans and the technologies which are required to extract them.

In the early seventies, the publication of *The Limits to Growth* (1972) and *The Energy Crisis* (1973) provoked widespread anxiety about the problem of finite resources and worries that these shortages might put a brake on economic growth. These concerns were most felt in the world's most affluent countries—in the United States, in Western Europe and in Japan, which is especially dependent on imports of oil and other raw materials. In this mood, the governments of the United States and Japan financed large research projects aimed at finding ways of extracting energy and minerals from the sea. Politicians talked of the ocean as the next frontier for man to penetrate after space.

Now, in the late 1980s, with oil at $9.00 a barrel and copper, zinc and other metals also at low prices, the enthusiasm has ebbed. Exploration budgets have been cut back even by the oil companies who are drilling in the North Sea. In this context, the researchers present a rather pessimistic view of the future for ocean research, though the position could change radically if, for some reason, we had another oil crisis or commodity crisis as we had in the early 1970s.

SCIENCE FICTION

The book still reads like science fiction. We read of *manganese nodules* the size and shape of potatoes which lie on the seabed and can be dredged by a continuous line bucket. They can be mined most conveniently in the Pacific Ocean, in areas like Hawaii or Mexico. The Japanese have a National Programme to spend $100 million on a mining test off Hawaii by 1990. The Western Mining Consortium, dominated by American companies, is concentrating its research in the Clarion-Clipperton zone in the north-east equatorial Pacific, probably hoping to bring the nodules into California or Mexico for processing. But what will the environmentalists have to say about the effect of dredging on marine life and the beaches of California?

Next we learn of attempts to recover *phosphorous nodules* from the continental shelf, and the project that aims to extract copper and zinc from the floor of the Red Sea. To raise the phosophorous nodules a simple suction pumping system may be sufficient. Or the mining companies may use a series of buckets on a 'conveyor belt'. The Union Oil Company was trying to mine phosphorous nodules off the coast of California—until they came across a number of unexploded naval shells which rendered further mining impossible. New Zealand has been exploring the area of Chatham Rise, which is within its 200-mile economic zone. But here too problems have been encountered—worries about the effect of scouring the continental shelf and worries about the disposal of dredged materials. These concerns have been dwarfed recently by the drastic effect of an abrupt fall in the world price of phosophorous.

The project to *extract metals from muds* on the floor of the Red Sea in the Atlantis 11 Deep is being financed by the governmment of Saudi Arabia and Sudan through a $20 million contract. Feasibility studies have been carried out by the West German firm Preussage and they have concluded that it should be feasible to pump the muds (with a consistency of toothpaste) from a depth of 2,000 metres. The future of the project is sensitive to the market price of silver, zinc and copper. This year it is hoped to run a full pilot mining operation to feed a land-based processing plant.

Next we come to *sands and gravels and radiolarian oozes*. Dredging for sand and gravel is the marine equivalent of open-cast mining and is very destructive of the marine environment—recovery of the sea floor is estimated to take two or three years but drifting of sediment clouds could have disastrous long-term consequences for fisheries. However, dredging will be essential if we are to build *artificial islands* in the North Sea and elsewhere. Radiolarian oozes are the skeletal remains of small marine animals. They have a high silicon content and could be a possible source of new light-weight ceramic materials with properties of chemical and heat resistance—which could be used, for example, in building mixed with concrete.

Then we read of *hydrothermal sulphides* which occur as rock containing silver and zinc around hot water vents in the sea-bed—for example in the Galapagos Rift Valley at a depth of two kilometres. To extract this type of rock will require a genuine deep mining ocean technology of a kind which does not exist at present. *Potash* is another useful mineral which occurs in very high concentrations in closed seas or lakes in hot climates—such as the Dead Sea. Extraction of potash for fertiliser has been going on for some years in the Dead Sea and the techniques employed there might be used elsewhere. *Uranium* can also be extracted from sea water by the use of ion exchange resins and the Japanese have a programme which plans for a plant to be built by 1990.

However, the most feasible projects seem to be based on shore-base stations. In 1982 the US government voted $75 million for OTEC research and $2 billion in the form of loan guarantees. However, under President Reagan the programme seems to have lapsed, though the US Department of Energy is still financing design proposals. There are some technical problems such as the building of a cold water pipe capable of carrying a rate of flow equal to that of the River Nile.

Ocean Thermal Energy Conversion (OTEC) has been described as the vanguard of research into renewable energy systems from the sea. It aims to use differences in temperature in the sea—which may be over 20° centigrade—to drive a turbine. Some schemes envisage the use of a Plant Ship which could process chemicals or metals. Or a floating platform station to feed electricity into the grid by a cable. The Japanese, however, are still in the field with a programme to build a land-based station this year or next.

Artificial islands are the ocean equivalent to space laboratories—new bases for research and manufacture. Airports, power stations, deep harbours—all can be prefabricated and floated out to sea. A Japanese consortium even ferried a complete operational pulp mill to Brazil! Now they have a project to construct a new industrial artificial island—the Kobe Port Island Scheme—complete with harbour facilities, residential, commercial and recreational areas.

The oceans are a mysterious and treacherous environment to work in and greatly improved techniques will be required for surveying, mapping and navigation if marine resources are to be located and exploited. Fortunately, a whole range of new and sophisticated techniques is becoming available. Acoustics and radio are the two basic technologies that are used. Acoustic methods, like echo-sounding and sonar scanning, are used in mapping the sea-bed and even the geology below the ocean floor. These methods are generally used with traditional magnetic and gravitational techniques, and with seismic reflection when deeper exploration is needed.

Acoustic, gravitational and seismic surveys are usually made from under the sea. Radio waves move above the water, operated either from dry land, from an aircraft or from a satellite. The advantage of satellite sensing comes from the possibility of covering large areas of ocean in a short time and producing a comprehensive picture for a reasonable price. Navigational and positioning services are provided by satellites in equatorial geostationary orbit at a height of about 36,000 km. or 23,000 miles. This feature is common to telecommunications satellites, which means that they can be provided as extra services by commercial operators who are mainly in the telephone business, whereas the much lower polar orbiting remote sensing satellites (at 500—1,000 km.) must be provided as a dedicated service. This means that satellite services for navigation and positioning are likely to become available earlier and more cheaply than satellite services for sensing.

Compared with remote sensing over land, the seas present considerable problems of interpretation, and ocean surveying by satellite is still at an early stage of develpoment. Nevertheless the first results are encouraging and several programmes are currently running. Remote sensing is effective in mapping major features such as sea temperature, ice and icebergs, levels of pollution and currents.

The nearest to a specifically marine resources satellite is the 'Landsat' series—an earth resources programme begun in the 1970s. Two further projects are planned for the late 1980s: the ERS-I project by a European/Canadian consortium and the Japanese Marine Observation Satellite. If these projects go ahead it is conceivable that they will provide such accurate estimates of resources for it to be possible for commercial investments in the new marine technologies to go ahead.

Ocean technology is clearly a new frontier which we are only just beginning to explore.

July 1986   Bernard Taylor
                                    Henley Management College

Preface

The account of the development of new marine technology reported in this book is based upon some aspects of the extensive research carried out by members of the Marine Resources Project at the University of Manchester. The Marine Resources Project was established under the direction of Glyn Ford to pursue research into the development of policy in respect of the future expansion of interest in marine technology. This book, however, concentrates on only one part of those studies; specifically those concerned with the mineral and energy resources of the oceans and the technologies which might permit the commercial development of those marine resources. Such studies, although by no means the exclusive interest of the Marine Resources Project, were nevertheless an important feature of its research programme between 1978 and 1984.

The Marine Resources Project completed studies of many schemes for new marine technology. The work has been in the form of literature surveys, summaries of previous marine technological developments and, primarily, original research concerned with the detailed analysis of the prospects for marine industry. The products of this research have been published as papers in the relevant journals, presented at conferences and collected as a series of working papers on new marine technology some of which have been, of necessity, restricted in their availability.

The authors of this book have sought to do no more than to present a review of the current state of new marine resource technology based principally on the studies by the Marine Resources Project's members during the six years that Glyn Ford was Co-ordinator. No new research is offered here for the first time. In order not to produce a book of only very specialist interest and in order to illuminate some of the themes common to a range of new technologies it has been necessary to reduce considerably the detailed content of the original reports and papers. It is intended that no serious errors have been perpetrated by this process, but if errors of simplification or interpretation have been made they are entirely the responsibility of the authors here. Any reader wishing to pursue these topics in greater depth is recommended to refer to the original publications which will provide more extended and detailed accounts as well as full references and bibliographic information. In order to facilitate this a select bibliography is included as an introduction to the literature on marine resource technology.

Acknowledgements

Full recognition must be given to the members of the Marine Resources Project at the University of Manchester whose work provided the basis of much that is reported here. Special thanks are due to Daniel Spagni, current Co-ordinator of the Marine Resources Project, and to Professor Michael Gibbons of the University of Manchester for their advice and co-operation and to Professor Bernard Taylor of Henley Management College. Appreciation must also be expressed for the efforts of the typists who have worked on various stages of the manuscript to render it legible.

The authors remain solely responsible for all errors of commission and omission. The views expressed herein are not necessarily those of the Marine Resources Project collectively.

1 Introduction

The oceans are a vast resource. They are, some say, the last great untapped resource on this planet. The oceans are a mineral resource. There are minerals deposited on and under the sea-floor and there are minerals dissolved in the body of the oceans' waters. The oceans are an energy resource. The continual flows of water as tides, waves and currents represent the immense potential energy stored in the oceans. It is energy that is renewed by the heat of the sun and the gravitational attraction of the moon. The oceans are also a resource of space. The surface of the oceans and the volume of the water mass are a resource for leisure, for farming marine crops and for siting installations on floating platforms or artificially created islands. Unfortunately, one of the ways in which the oceans are perceived as a resource of space is as a sump into which all manner of waste can be dumped, trusting to dilution and dispersal to render it harmless. But recently the possibility of utilizing the oceans' other resources has been considered more seriously.

The existence of minerals in the sea has been known for a long time and the oceanographic surveys of the middle and late nineteenth century greatly increased knowledge of mineral deposits, particularly in the deep oceans. But they remained a scientific curiosity and were not considered to be of commercial importance until in the years after the Second World War when the possibility of the depletion of non-renewable resources limiting future industrial growth in the West became a matter for concern. Several studies during the 1960s[1] suggested that economic growth might be stifled and serious social and political consequences might follow, if non-renewable resources continued to be consumed at the rates that had prevailed in the two decades of post-war reconstruction. New sources of supplies were therefore sought and the search for renewable sources became a particular focus of attention. The mineral resources of the oceans became a principal target in that search as these were no longer seen merely as a scientific curiosity.

Even if the direst predictions of resource depletion were felt to be exaggerated, it became generally accepted that certain industrially important materials might well be subject to severe price rises within two or three decades, as more easily-won reserves[2] were exhausted and new supplies would involve greatly increased production costs. Therefore, in this case also, it would be prudent to investigate the possibility of the oceans supplying minerals which, in the long term, could be cheaper than obtaining difficult and expensive land-based resources. Such considerations helped to stimulate interest in the option to obtain minerals of marine origin.

The range of minerals in the oceans is large, although the concentrations are very low in some cases. Others, however, do occur in higher concentrations. For example, on the continental shelves there are extensive deposits of sands and gravels as well as minerals, containing phosphorus, barium, iron, chromium, titanium and zircon. On the slopes between continental shelf and the deep ocean there are ferromanganese oxides, phosphorite nodules and muds containing high concentrations of zinc, copper, lead and silver. In the deepest oceans (below 3,000 metres) there are polymetallic nodules containing iron, manganese, copper, nickel and cobalt, and there are deposits of organic muds such as calcareous and siliceous oozes. There are also huge deposits of minerals beneath the sea-floor. Apart from oil and gas, below the continental shelf there are extensive measures of coal and sulphur and below the floor of the deep oceans there are large deposits of rich metallic sulphides. In the body of the seas there are traces of many metals and salts in solution, including magnesium, calcium, sodium, bromine, potassium and uranium. There is also dissolved gold, but its concentration is low, even compared to the other trace metals present.

As has already been said, minerals are not the only resource of the seas. The energy of the oceans is enormous. It is found not only in the kinetic energy of waves, currents and tides, but also in the gradients of temperature, salinity and density that exist within bodies of water. If that energy can be harnessed (probably for the generation of electricity) then a renewable supply of power will be available that will replace dwindling fossil fuel reserves and provide a long-term alternative to nuclear power. But, analogous to the low concentration of minerals in the ocean, the energy in the ocean is thermodynamically low-grade. In effect this means that although the energy there is renewable, and of itself cheap, it is not in a form easily suited to the requirements of large-scale electricity generation.

It is not particularly useful to speculate as to the increase that marine minerals and ocean energy can provide in world resources, except to realize that although the theoretical amounts may be considerable it is far from trivial to turn those theoretical resources into viable sources of supply. It is that transformation with which this book is concerned. The book will look at the novel (but not always new) technologies by which the development of the ocean's resources may be accomplished and will take the form of a series of case studies of proposals for new marine resource technologies. These proposals cover a wide range of schemes from the simple to the exotic and from current projects to ones that are still at least half a century distant. Chapter 2 deals with the mining of polymetallic nodules from the deep ocean. Chapter 3 looks at the rather less ambitious schemes to raise metal-rich muds from the floor of the Red Sea and to dredge for phosphorite nodules. Chapter 4 then reports on the progress of several other marine mineral proposals, including some that are currently only in the earliest stages of development. Chapters 5 and 6 turn to proposals for generating power from the oceans. The former is given over to a detailed study of ocean thermal energy conversion systems (OTEC), which received extensive support in the United States during the 1970s, and the latter reviews other schemes, which range from the tidal power schemes already in operation through to proposals for generating electricity from the salinity and density gradients in the oceans. Chapter 7 reviews developments in the construction of artificial islands and the uses to which they could be put. In particular, artificial islands, both fixed and floating, offer the prospect of increasing the area for industrial expansion in crowded coastal locations and for removing certain industrial operations from areas of population. Chapter 8 differs slightly from the format of other chapters in that it deals, not with a specific project of marine resource development, but with a technology which will be important in enabling progress to be made across a range of marine programmes. It is concerned with the search for ever more sensitive and accurate techniques of surveying the oceans. In particular the chapter considers the prospects for surveying from satellites. The importance of these techniques is that the higher the quality of the initial general surveys of the oceans the more efficient and specific can be the detailed surveys at the surface. They are, therefore, relevant to many individual schemes. Chapter 9 summarizes the evidence from the case studies and attempts to assess the prospects for new marine technologies in the next few decades.

Not included in this study are fishing, shipping and the offshore oil and gas industry. This is not because there will not be new and important developments in those industries, but because that emphasis in this book is upon the development of *novel* technologies rather than upon the continued development of existing ones. Indeed fishing, shipping and offshore oil and gas production will not be entirely excluded as developments in those areas may be relevant to the new resource technologies. For instance, advances in the use of satellites for navigation are primarily intended for shipping, but they will be very important in permitting the necessary accuracy of station-keeping for surface vessels at deep ocean mine sites. Similarly developments undertaken in the construction and servicing of offshore installations in the oil and gas industry could be applied to artificial islands in general and could permit savings of time and money in their development. But, these spin-offs notwithstanding, the future of more traditional marine activities is not a central theme of this book.[3]

Each case study is not confined to the technological means of resource exploitation. The technical challenges associated with the extraction of dissolved minerals from sea-water or with the retrieval of nodules from a sea-bed several kilometres below the surface are not trivial. Considerable imagination and technical skill will be required for these projects to be realized. But the solving of technical problems is not necessarily sufficient to ensure an operational future. Obviously, the commercial prospects for the projects are important. If they are to be run commercially they must be able to produce the goods at a price which will find a market and still show enough return on investment to attract finance. Yet, as the studies will show, these schemes are rarely subject to commercial factors alone. There is a strong political component that cannot be ignored. The political dimension is expressed in many ways. It is seen in the extent of government support for projects; this support may be financial or it may be in the form of enabling legislation. It is also seen in the international context of marine technology which raises several questions about the conduct of marine projects in international waters. In particular, it implies the need to address questions of the ownership of the mineral, and other, resources of the deep oceans and of the responsibility for the environment of the open seas. Each case study is therefore assessed not only with respect to the technical and commercial aspects, but also within the framework of domestic political attitudes and the framework of international legal considerations.

Few of the novel marine technologies are likely to be sited close to Great Britain. Yet a study of the recent history and future prospects of these technologies is not irrelevant to firms in that country, since many companies (in both Britain and other countries of North-West Europe) have gained valuable experience from two decades of oil and gas production in the North Sea. The importance of that experience should not be underestimated since it has provided the opportunity to test techniques and materials in a harsh marine environment. Such a proving-ground has not been available to firms elsewhere and companies that have operated in the North Sea ought to be well placed when choosing to seek involvement in new marine projects away from North-West Europe. However, in this book there is no intention to offer any simple answers to the questions of commercial participation in new technologies; the aim is to begin to define the context within which the commercial decisions will have to be made.

The case studies will show that in the recent past the high point of enthusiasm for new marine technology was in the mid-1970s, when several factors came together to promote marine technology as the vanguard of American high technological enterprise in the years after the end of the Apollo space programme. The aim of the enterprise was to ensure American economic growth and industrial power by ensuring, as far as possible, self-sufficiency in the supply of energy and raw materials regardless of scarcity or political disruption of traditional sources of supply. The case studies demonstrate that since the mid-1970s conditions have changed. Those factors that weighed so heavily in favour of marine technology are no longer so important. The scale of R & D has ceased to grow as rapidly as before and many projects seem to be marking time. It is tempting to suggest that marine technology has had its day; that it was appropriate to the United States at that time, but is no longer in keeping with the political and economic climate of the 1980s. But the case studies also show that to be an over-reaction. The projects of novel marine technology can be adapted to the new conditions and indeed some have already done this. Although the prospects are not as spectacular as in the scenarios of ten or fifteen years ago, there remain excellent opportunities for marine technologies to develop within the conditions which now prevail.

Far from interpreting the current situation as indicating the demise of hopes for novel marine technology, there is now evidence of a wider interest in its future. The studies illustrate the fact that participation is no longer confined to a small number of American high technology corporations. Instead, many countries, most notably, but not exclusively, Japan, are now expanding their participation. This diversified base augers a sounder future and steadier development than that predicated solely upon the American programmes of the mid-1970s since there is not the reliance upon the commitment of a single country nor yet even of a single corporation. Therefore, rather than discounting marine technology as a phase now passed, it is timely to reassess its future and to see how it might develop, and even flourish, in the new circumstances that it faces for the remainder of this century.

NOTES

1. The most famous of these studies was D. H. Meadows *et al.*, *The Limits to Growth* (New York, 1972) which predicted economic and ecological catastrophe if the rate of consumption of natural resources continued to increase unrestrained. The assumptions of the computer modelling in the Meadows study were subject to much discussion and some critics argued that the models were insufficiently sophisticated for simulating anything as complex as the world economy. Nevertheless, *The Limits to Growth* was very influential in the early 1970s because it showed that the problem of finite natural resources was not one that could be ignored.
2. In discussions of raw material economics, 'reserves' and 'resources' are usually not synonymous. Generally, 'resources' refer to the total amount of a raw material identified as existing, whereas 'reserves' refer only to that portion of the total which are currently economically viable. As this book is concerned with technologies of which the economic viability is unknown, it is aimed

primarily at the question of resources, which may prove subsequently to be viable reserves.

3. It should be noted that this study is concerned with civil programmes for new marine technologies. However, it is acknowledged that some schemes may have a military component and it should not be forgotten that an interest by the military often has the consequence of securing finance for R & D which might not otherwise be forthcoming. Nevertheless the projects here are assessed entirely as civil ones.

2 Manganese nodules

INTRODUCTION: NODULES AS A RESOURCE

One of the earliest extensive oceanographic surveys was made by the British 'Challenger Expedition' between 1873 and 1876 and one of its discoveries was that the deep ocean was littered with small, dark brown mineral accretions. On analysis, these were shown to be primarily composed of iron and manganese, together with smaller amounts of other metals. As such, 'manganese nodules' became, and remained for some eighty years, a mineralogical curiosity. However, during the last three decades, interest has been increasingly focused on the potential of manganese nodules as a commercially viable source of several metals.

The nodules are typically the size and shape of potatoes, being 6 to 8 cm. long along the largest axis. They lie exposed on the surface of the sea-floor, at depths of around 5 km. and, although confined to low and middle latitudes, they have been found extensively distributed within those limits. Nodules may contain over twenty metallic elements and the proportion of each varies, but an average one, after all water content has been removed, is 25 to 35 per cent manganese, 6 per cent iron, 1 to 2 per cent of both nickel and copper, 0.1 to 0.5 per cent cobalt and between 0.001 and 0.01 per cent molybdenum and vanadium. Smaller traces of other metals have been found.

When the interest in nodules as a commercial proposition was revived in the late 1950s, it was sustained by the hope that a bonanza of minerals was waiting, ready for exploitation, in the oceans. The subsequent history of nodule mining has been considerably more complicated than the early projections assumed. There have been many obstacles, not only financial and technical, to commercialization. In this chapter some of these factors will be considered in an attempt to assess the current state and future prospects of this mineral resource.

The mechanism whereby manganese nodules (also called polymetallic nodules or ferromanganese nodules) are formed is still subject to dispute. There is consensus, however, that their occurrence corresponds to those areas of the sea-floor subject to very low rates of sedimentation. This tends to favour the deep oceans since they are usually the most remote from land and do not experience sedimentation originating from the continental shelf. Furthermore, there is the influence of the 'carbonate compensation depth'. This is the depth in the ocean below which calcium carbonate, which is the mineral found in the skeletons of many small organisms, becomes soluble in water as a consequence of temperature, pressure and the oxygenation of the water. Hence, where the sea-floor is below this level (typically 4 to 5 kms deep), carbonate minerals, whether derived from the weathering of surface limestone or from organisms, will not reach the sea-bed as sediment. This again leads to low sedimentation rates being associated with the deep ocean.[1]

The distribution of nodule deposits in the lower and middle latitudes is also related to slow sedimentation rates. In these latitudes small marine organisms are predominantly calcareous, while in higher latitudes organisms with skeletons of silicon minerals predominate. There is no corresponding compensation depth for silicon minerals below which they will dissolve. Consequently, regardless of depth, the sea-floor in higher latitudes is continually subjected to sedimentation from the skeletal remains of siliceous organisms.

The fact of the geographical distribution of manganese nodules, together with the historical location of earlier investigations, has tended to concentrate recent research into the commercial prospects of nodule mining on an area of the north-east equatorial Pacific called the Clarion–Clipperton Zone. This emphasis on a comparatively small piece of ocean has meant that even in the late 1970s there was no reliable extensive survey of nodule resources over wider tracts of the sea-bed. Without such surveys it is impossible to begin to assess the viability of nodule mining on a large scale; nevertheless, commercial interest has already been very serious even though it is now accepted that the earliest predictions of a bonanza were wildly optimistic.

COMMERCIAL INTERESTS

Despite the uncertainties, as early as 1959, several companies investigated the potential of nodule mining. The preferred organization of such exploration has been the consortia where, in principle, the special skills of individual firms can be combined to pursue the new enterprise and where, again in principle, both the commitment of resources and the risks can be shared. The membership of the consortia is always open to change and some of the groups have been more active than others, but at the beginning of the 1980s there were five consortia still operating (see Table 2.1).

Although European companies are well represented in the consortia it is the corporations from North America which dominate, both in their numbers and their percentage holdings, within the consortia, and hence are effectively the major partners. This reflects the fact that the consortia were formed primarily at the instigation of the American firms, which for political reasons desired some element of international co-operation and

Table 2.1 Consortia involved in ocean mining: 1980*

	%
Kennecott	
Kennecott Copper Corporation (USA)	40
Rio Tinto Zinc (UK)	12
Consolidated Goldfields (UK)	12
BP Minerals (UK)	12
Noranda (Canada)	12
Mitsubishi (Japan)	12
Ocean Mining Associates (OMA)	
US Steel (USA)	33
Sun Oil (USA)	33
Union Minière (Belgium)	33
with Deep sea Ventures (USA) as engineering contractors.	
Ocean Minerals Company (OMCO)	
Ocean Minerals Inc:	
Royal Dutch Shell (Billiton) (Netherlands)	25
Bos Kalis Westminster (Netherlands)	10
Lockheed Missiles (USA)	40
Standard Oil of Indiana (Amoco Minerals) (USA)	25
Ocean Management Inc. (OMI)	
International Nickel Co. (INCO) (Canada)	25
AMR (Preussag, Metallgesellschaft, Deutsche Schachtbau) (Germany)	25
DOMCO (23 companies headed by Sumitomo) (Japan)	25
SEDCO (USA)	25
AFERNOD (France)†	
Bureau de Recherches Géologiques Minières (BRGM)	
Centre National pour l'Exploitation des Océans (CNEXO)	
Commissariat l'Energie Atomique (CEA)	
Chantiers de France—Dunkerque	
Société Minière le Nickel (SLN/SMN)	

* Three other groups exist: the Continuous Line Bucket Group (consisting of 19 companies, predominantly US and Canadian); Deep Ocean Mining Association (32 Japanese companies); Eurocean (4 companies; 3 Swedish, one French). There is little evidence that these groups are sustaining a continuing effort at present.

† No share-holding percentages are available for this group—most publicly owned.

approached European firms to join them. The criterion for membership of consortia should be that the consortia will achieve the blend of expertise necessary to develop the new resource. This blend would probably be a combination of skills that no individual firm could provide. For example, the Ocean Minerals Company consortium is comprised of Lockheed, which is a high technology firm experienced in advanced innovative projects, two oil companies with experience of shipping and marine operations that wished to diversify into other natural resources, and Bos Kalis, which is a marine engineering company. This appears to be an ideal combination of different specialities. Furthermore, there are two American firms and two European companies, with the Americans holding 65 per cent of the consortium's shares. Ocean Minerals Company is a typical mixture, both in the nationality and the expertise of its participating firms. Needless to say, the practice of mutually beneficial international collaboration is often less impressive than the rhetoric.

Each consortium proposed similar two-phase programmes of R & D to precede commmercial mining. Typically, 'Phase 1' objectives include detailed sea-bed surveys, mine-site evaluation, equipment design and prototype testing. During the early 1980s, most consortia had completed the bulk of their Phase 1 projects and each had spent in the region of $50 million up to that point. 'Phase 2' was to cost around $150–200 million over two or three years and to involve the construction of the first commercial-scale plant for lifting and handling the nodules, though not the construction of the process plant to refine the nodule ore into the desired elements. The lifting apparatus has to demonstrate its operation on a commercial scale before the extra resources will be committed to build the processing plant. In fact, there has been a hiatus since the completion of Phase 1 development. The consortia have been reluctant to proceed until there was some evidence that the demand for (and hence the price of) nodule metals was beginning to rise again and until the establishment of a legal and political framework for deep ocean mining had been assured. Of both these issues, more will be said below, but until such time as they are clarified, the consortia will proceed very cautiously. The bulk of investment by the consortia has been privately financed. The North American firms, for instance, have probably received federal funding only to finance the environmental research into the impact of nodule mining as required of them by United States law. The French and West German companies are more likely to have received grants towards R & D expenditure, while in Britain the government has supported British firms' participation in the consortia but it has not contributed towards R & D, most of which has been undertaken in the United States. This all reinforces the overall situation in which the development of nodule mining is predominantly an American enterprise in which European firms are participating.

Since the European programmes are so closely associated with American effort, the only other major independent programme is that of Japan. Mitsubishi and the DOMCO group of companies are members of international consortia, but Japanese interest goes beyond that alone. Japan's shortage of indigenous mineral resources leads it to adopt a very positive attitude towards the investigation of possible new sources of raw materials. To that end the government has supported nodule mining development since 1969, when the research vessel *Hakura Maru No. 1* was commissioned for sea-bed surveys. Subsequently, a forum for government and industrial collaboration was created in 1974 through Deep Ocean Minerals Association, with a membership that has varied between thirty-three and thirty-nine companies, but which in the later 1970s was apparently doing little more than surveying.

The next important stage in the Japanese programme came in 1981 when nodule mining was classified as a 'Large Scale Project' under the 'National Research and Development Programme'. This denotes those projects of great national importance, the risks and expenses of which are so great as to discourage private industry from proceeding at the pace the government would prefer. The status within the National Programme also implies a degree of official secrecy that makes it difficult to assess Japanese progress, but the current programme aims to spend $100 million by the end of the decade and to achieve a one-quarter to one-third scale mining test off Hawaii by 1990. Although they began somewhat later than the Western consortia, the Japanese have been expanding their programme at just that time when Western interest has been tending to progress more slowly.

Japanese development will remain primarily financed and co-ordinated by government, but it is not intended to become a state-run industry. After the conclusion of successful mining tests, the government intends to withdraw and the companies will be expected to continue the drive to commercialization and to compete with the other international consortia thereafter.

MINING SYSTEMS, TRANSPORT AND PROCESSING

Polymetallic nodules are no more than an alternative to other sources of the metals they contain. Even after due allowance has been made for strategic considerations of mineral supply metals derived from nodules must be produced at a cost comparable to conventional ones. The development of a viable and economic system of nodule mining and refining is essential to the prospects for their commercial exploitation.

There will be three phases to the process (see Figure 2.1). In the first place the nodules must be collected from the sea-floor and subsequently conveyed to the mining vessel on the surface. This phase requires the greatest technological development and involves the greatest risks to the overall technical feasibility of metal extraction from nodules. Financially, however, recovery of nodules from the sea-floor will probably only account for some 20 per cent of both capital and development costs. The second phase is the transportation of the nodules from the mine-site to a

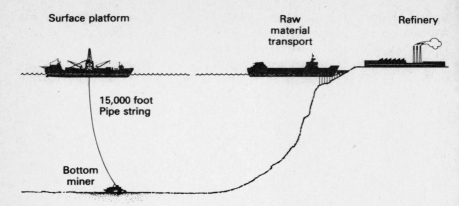

Figure 2.1 Manganese nodule mining and processing system. *Source*: Ocean Minerals Company.

refinery on shore. Technically, this courts the least risks as bulk handling and transportation of ore by sea is common practice in the mineral industry. Estimates suggest that these operations would take 10 per cent of capital and 15 per cent of operational costs. The final phase is processing the nodules into metals. It is considered of only intermediate risk technically. The conventional engineering of mineral refining provides a well-tested framework within which the specific problems of nodule processing should be relatively easily solved. Initially, it will be undertaken in processing plants on land and it will comprise the most expensive part of the operation, with 70 per cent of capital costs and 65 per cent of running costs likely to be consumed by it.

This creates an interesting situation. The technically difficult phase, that is, the recovery of nodules from the sea-bed, only accounts for a small proportion of overall costs. If the cost of developing or operating nodule collection were to be double the current estimates, it would only add 20 per cent to the total cost, whereas a much smaller percentage increase in the cost of processing could increase overall expenditure by considerably more than 20 per cent. This is somewhat comforting for the consortia since the costs of the project are not unduly sensitive to rising costs in the high-risk R & D areas.

NODULE COLLECTION

The system by which nodules are collected and conveyed to the surface vessel has already been the subject of extensive investigation and investment. Two classes of system have been proposed. The first is the Continuous Line Bucket (CLB) (see Figure 2.2), which is a dredging arrangement of buckets attached to a very strong loop of cable up to 15 km. in length. The buckets are dragged along the sea-bed, where they collect the nodules and are then drawn up to the surface to be emptied. It is technically unsophisticated and inefficient but it is also relatively cheap. Early tests with one ship have proved that there are difficulties in controlling the cable, although a system stringing the loop between two ships seems to be better. Two main problems remain: ensuring the buckets collect a reasonable quantity of nodules and ensuring the cable is strong enough to take the strain. The buckets are not controlled and may not pick up any nodules as they are dragged around the loop, thus making it difficult to guarantee an efficient sweep. If a 15 km. cable were used, it would need to be a steel cable significantly stronger than those currently available for marine uses.

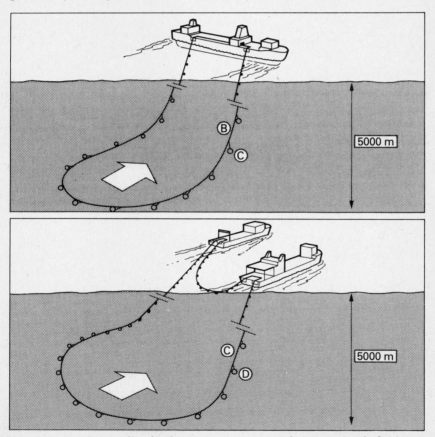

Figure 2.2 Continuous line bucket system. *Source*: Arbeitsgemeinschaft Meerestechnish Gewinnbare Rohstoffe.

CLB systems might be profitable for small-scale operations lifting between 150,000 and 200,000 tonnes of nodules per year, but they are quite unsuitable for projects of the scale envisaged by international consortia. Although interest in CLB dredging is periodically reawakened, it is unlikely to figure much in the future and no consortium is currently pursuing it. The consortia have all opted for the second category of collection system, which is one that uses an hydraulic lift. There are two variations on this. The first is the 'air-lift', where compressed air is injected into a pipe that runs from sea-bed to surface vessel. The buoyancy effect of the rising air lifts the nodules. The second type is a pumped system in which an electrical pump (or pumps) inside the pipe provides the lifting force.

Whichever hydraulic lift system is selected, the quantity of material to be raised implies very high flow-rates within the pipe. Typically, a commercial mining programme would need to raise 3 million tonnes of nodules per year. If, after due allowance for failures, servicing and any other interruptions to mining, it is possible to operate on three hundred days in a year, then a flow-rate of 10,000 tonnes of nodules each day is necessary to meet a 3 million-tonnes target. The total material lifted will be greater than 10,000 tonnes a day, of course, because raising that weight of nodules will undoubtedly also raise sediment, however efficient the filtering may be.

The choice of lifting system must balance several considerations. The in-line electric pumps are less energy-intensive than the air-lift option because of the high energy consumption involved in pumping compressed air down to a depth of several kilometres. The electric pumps would reduce the energy costs of lifting nodules. Unfortunately, they too suffer one major problem. There will be severe wear on the impeller blades of an in-line pump, owing to the high flow-rate of nodules within the pipe. Furthermore, the worst wear will occur on the deepest and least accessible pump. Servicing and repairing will be difficult and expensive, so both from the point of view of total running costs and because of the need to achieve uninterrupted mining, it is essential that in-line pumps are highly reliable. Nevertheless, in-line electric pumping is currently favoured, provided suitable strengthening can reduce the rate of wear on the rotors.

The efficiency of the collector's operation will have a great impact on the overall costs of raising nodules. The design of the dredgehead is therefore crucial. It must operate on undulating terrain, which has only small surface irregularities, maintaining sufficient contact with the sea-floor to achieve efficient nodule collection, while not lifting unwanted materials such as sediment. It is necessary to balance the increased width of a collector that will make broad sweeps against the decreased contact a wide dredgehead would have with the contours of the sea-bed. Although a narrow dredgehead could follow the undulations closely, it would need many more traverses and a greater forward speed in order to compensate for the narrower sweep made and, since the first generation of collectors will be passive (i.e. pulled by the surface vessel), there is a limit to the speed at which a collector can move. The first dredgeheads will therefore be fairly wide, even if this reduces the efficiency at which the terrain is cleared of nodules. Collectors of up to 40 m. in width have been tested, but about half that width seems to give the best balance between the area covered and the efficiency of collection for a passive dredgehead. The addition of a screening system to exclude unwanted material from being lifted will be important since any sediment lifted will increase the costs for each unit of nodules raised. It will be no use designing a dredgehead that efficiently recovers every particle of mud and sand as well as the nodules, so a simple filtering system that retains the larger nodules but allows sediment to escape will be essential.

Later designs of collector may not require to be pulled by the surface vessel. An active collector has the benefit that it can 'graze' an area of sea-bed under its own motive power while the surface vessel holds station above the mine-site. Furthermore, an active dredgehead will not be restricted to the forward speed of a towing vessel (see Figure 2.3). Other refinements in second-generation dredgeheads would include better screening, to increase the efficiency of the lift, and some pre-crushing of the nodules prior to lifting, since it is preferable to pump a homogeneous material rather than discrete lumps of rock. However, such sophistication in the most vulnerable and inaccessible component of the mining system is unsuitable for the early commercial programmes and so a passive dredge-head is the initially favoured mode.

Whatever combination of collector and lifting system is selected, the nodules, together with any other material that escapes screening, will be conveyed to the surface in a pipe. Even to transport 10,000 tonnes per day, the pipe requires a diameter of only 1.5 m.; it will, however, be more than 5 km. long! It is unlikely to be made of steel and, despite its length, the use of screw connecting sections and techniques from conventional drilling practice render the construction and deployment of the pipe relatively unproblematic. However, it is the dynamic behaviour of such a pipe in the water that causes concern. It will be subject to stresses from the relative movements of dredgehead and surface vessel as well as to stresses due to the currents of the water masses through which it passes. The design of the pipe must accommodate these forces in order to produce a structure that is flexible in use and resistant to fatigue and failure. There will also be vibrations within the pipe resulting from the flow of material through it. They, too, could contribute to fatigue, but the maintenance of a constant rate of flow within the pipe will reduce the risk. Hence, whether in-line pumping or air-lift is employed, it will be through multiple units placed at

Right: Power Driven
('Active') Collector
Below: Towed Sled
('Passive') Collector

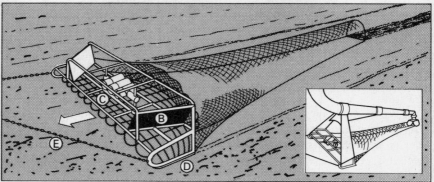

Figure 2.3 Types of nodule collector. *Source*: Arbeitsgemeinschaft Meerestechnisch Gewinnbare Rohstoffe.

intervals along the pipe in order to achieve a constant pressure head, and so reduce vibrational stress.

The final component of the mining system is the surface vessel, which is the single largest capital item in that sector, constituting over half of estimated expenditure. The vessel must be capable of precise navigation, both to and at the mine-site. It must carry and deploy the dredgehead and pipe and it must also have capacity for the storage and transfer of recovered nodules consequent upon mining rates of 10,000 tonnes per day. This points to a commercial size of around 100,000 tonnes dead weight. Beside these mining requirements, the vessel must provide crew facilities comparable to the standards on oil-rig accommodation platforms.

The stability of the vessel is fundamentally important. Not only will good stability permit mining to continue longer in heavy seas than would otherwise be possible, but it can also reduce the stresses on the pipe caused by the roll, pitch and heave of the vessel in ocean swell. For the longer term, semi-submersible vessels offer much better stability than surface vessels, but initially most consortia made tests with conventional deep sea drilling ships as surface vessels. These ships already have proven deep sea capabilities and, with the addition of heave, roll and pitch compensation mechanisms, they provide a satisfactory basis for testing at sea.

For commercial operations, typical designs postulate an open well within the vessel's hull through which mining apparatus would be raised and lowered. This configuration protects the equipment, in heavy seas especially, and when deployed within a pool of relatively calm water. The collector and pipe would be supported from a hydraulically operated, gimballed derrick positioned over the well.

Accurate navigation and precise station-keeping capabilities will probably be achieved by the use of navigational satellites to determine an exact position and then variable thrusters to maintain that position at the mine-site. Many useful techniques are now being developed for the off-shore oil industry and their transfer to nodule mining may be anticipated.

NODULE TRANSFER AND TRANSPORTATION

This phase poses the least problems. Both in capital and operating costs, it represents only a small part of total expenditure and the technical difficulties are not severe since bulk ore transportation is a well-tried operation. Nevertheless, the sheer quantity of nodules that will have to be transferred at sea from the mining vessel to the bulk carrier does present some challenges for those concerned. Transfer must be continued, even in adverse weather, since storage capacity on the mining vessel will be strictly limited and it would be disastrous if mining had to be halted until calmer conditions permitted nodule transfer to resume. Mechanical conveyance of whole nodules may be more susceptible to interruptions from the weather than a hydraulic system pumping a slurry of crushed nodules. This requires crushing the nodules on board the mining vessel;[2] however, apart from drying the nodules (to avoid transporting water), the first generation of mining vessels are not intended to undertake any processing. However, crushing may be carried out at the mine-site for ease of transportation. This goes against the conventional wisdom which asserts that, wherever possible, material should be transported in the most refined state possible. In the early stages the surface vessels will not be suitable for undertaking large-scale processing and so the ore will be shipped in a raw, though possibly crushed, state.

PROCESSING

The task of refining the nodules to extract the metals is the single largest consumer of both capital and current expenditure. The choice and design of an appropriate extractive metallurgy are subject to the well-tried criteria of chemical engineering practice and therefore the major problem seems to be more concerned with the selection of a process.

The choice of extractive process depends primarily upon which of the metals from the nodules are to be refined. Ironically, manganese itself is not necessarily a target metal; indeed, only one consortium is currently intending to extract it. All the consortia are agreed on the extraction of nickel, copper and cobalt. Unless the world demand for manganese increases dramatically (and currently it is almost entirely used in steel-making), the quantities of manganese produced by one consortium operating a 'four metal' extraction will be sufficient to meet demand. It is unlikely, therefore, that more than one consortium will go ahead and extract manganese.

There are two alternative extractive processes: pyrometallurgy (smelting) and hydrometallurgy (solvent extraction). There are few options within the former, but several different leaching treatments can be devised, depending on the selection of solvent. Leaching by sulphuric acid, hydrochloric acid or ammonia are all feasible. Each has its own technical advantages and drawbacks which each consortium must balance in making its choice.

Both pyrometallurgic and hydrometallurgic processes will extract around 90 per cent of the nickel. Copper recovery from the former would be 80 per cent, but about 90 per cent again in the latter process. Pyrometallurgy, however, will achieve a higher cobalt yield than solvent extraction does. Manganese is more straightforwardly extracted by pyrometallurgical means, but as only one consortium has targeted that metal this will not be a factor in the choice for other consortia.

Solvent extraction plant appears to be cheaper to build and operate and, since the extraction of manganese is not a major concern, current preference favours a hydrometallurgical process. It will probably be ammonia-based for the extraction of copper, nickel and cobalt. But final choices do not need to be made yet and all parties will delay before committing resources to a specific plant and process of a commercial size until full-scale mining systems have been proved.

The size proposed for commercial processing would entail a plant capable of servicing the ore from two mining operations (i.e. some 20,000 tonnes of ore per day). This is not excessive by the standards of mineral engineering, but the plant must continue to operate to its specified design performance, since the overall cost of metals from nodule ore is highly sensitive to the cost of processing. So, even if the novel challenges of marine mining technology are solved, the commercial future of nodules depends critically on the ability of the consortia to produce metals at competitive prices.

ECONOMICS: MARKET AND DEMAND

While strategic factors cannot be ignored in questions of resource supply,[3] it is unlikely that nodules will become a major source of copper, nickel and cobalt unless they are produced at prices comparable to traditional sources. There must also be sufficient demand to utilize the new supply of metals and to prevent market prices from being depressed by excess supply. It is difficult to predict the market for and the price of metals in, say, fifteen years time. However, the US Bureau of Mines (USBM) has produced a range of forecasts of demand for the four major nodule metals until the end of this century, based on different rates of growth for the industrial economies. Even using the lower growth estimates (which do reflect current growth-rates), the USBM figures suggest an increase of 2 to 3 per cent per annum in demand for the metals. Hence, providing that costs can be competitive with land-based production, there should be an expanding demand for the metals of nodule origin. The rate at which ocean mine-sites can enter production may be limited, however, since the output of one ocean mine, raising 3 million tonnes of nodules each year, would produce sufficient manganese and copper to meet one year's increase in world demand, and the nickel and cobalt from a single mine would be equivalent to two years' increment. If too many ocean mines came into production together, it could cause a glut and depress market prices. As long as this is avoided, the market should be able to absorb the new supplies without the prices of metals being reduced.

Even if there is a market for metals of nodule origin, those metals must still be competitively priced. The highly unknown factor in assessing the likely cost of nodule metals is the quality of the ore. A commercial grade of nodule will need to have a combined nickel/copper content of 2.25 to 2.4 per cent. If nodules of this composition achieved an average distribution of at least 10 kilogrammes per square metre over a total area of 50,000 square kilometres, then the site would be suitable to support a single mining operation with at least twenty years of active life. Furthermore, of course, the sea-bed's topology will determine what proportion of the nodules can be retrieved; retrieval rates of 50 per cent would be excellent.

The costs of mining, transportation and processing have been estimated by the consortia. One imponderable remains the mine-site itself. It must contain high-grade nodules, of high density, over a large area of sea-floor, and that sea-floor must be suitable for nodule retrieval. Hence, surveying has played a very important part in the early programmes of all consortia. The most crucial decision for each consortium that goes ahead with mining will be the selection of a mine-site. If the choice is wrong there is little chance of commercial success, regardless of how well the rest of the mining operation might perform.

Less critical than either the question of world demand for metals or site selection, but still commercially significant, is the location of the processing plant. Since, contrary to conventional practice, initially at least, unprocessed nodules will be shipped from site to the plant, it will be advisable to locate the plant as close as possible to the mine-site. If early mining operations are sited in the Clarion–Clipperton Zone, as seems probable, the coast of California will be the obvious choice for processing. Not only is it close to the mining, but it is also an industrialized region offering the necessary infrastructure of electrical power supply and good communication to industrial markets for the refined products. This is in contrast to the other possible locations, such as Mexico and Hawaii, where the infrastructure of industrial services is less developed. (If the provision of those services were to be borne in the cost of construction of the processing plant, the increase in capital costs would lead to higher-cost products.) However, California is the most environmentally conscious state in America and it may be difficult for a new mining industry to meet state requirements, especially on water discharge and the disposal of exhausted tailings.[4] Whether these problems will be severe enough to force consortia to locate plant in another area remains to be seen.

In summary, there are several criteria to be met if nodule metals are to have a chance of commercial success. The mine-sites must be of high-grade ore over a large area of sea-floor; the lifting, transportation and, most crucially, the processing sectors must all operate to specification; and there must be sufficient world demand for the metals thus produced.

LAW OF THE SEA

Nothing in the foregoing suggests insuperable barriers to the exploitation of nodule resources. There are technical challenges (which are fortunately in the relatively cheaper phases of the operation), there are the problems of scaling-up from pilot plant to commercial scale and there is the need for a suitable commercial climate. But in these alone, nodule mining hardly differs from many other large industrial projects. Moreover, since the early 1980s, the prospects for progress have been quite encouraging, both technically and commercially. Yet all the consortia (and even the Japanese state programmes, to some extent) have remained cautious about embarking on the next phase towards commercialization and herein lies one of the most important issues both for nodule mining and for the future of other new marine technologies.

Nodule recovery will take place on the floor of the deep oceans in what used to be considered 'international waters', so the question arises: whose nodules are they? This debate has mainly taken place within the forum of the Third United Nations Conference on the Law of the Sea (UNCLOS III) during the later 1970s and early 1980s, where the developing nations have sought to restrain the industrialized ones from embarking on unregulated ocean mining. UNCLOS III was concerned with many wider issues, but nodule mining served to focus attention on the immediate question of the physical resources of the seas.

The precise political alliances at the conference were complex, but, broadly speaking, the developing, non-aligned nations (known as the 'Group of 77') believed that bonanza profits were available to the consortia and they were concerned not to be excluded from the benefits of resources which they claimed were a common heritage of the deep oceans, outside national economic zones. The prospect of huge profits has now receded somewhat, however, and much of the proceedings at UNCLOS III must be seen as part of the more general debate over resource exploitation and the economic relationship between North and South. Furthermore, the Group of 77 included countries which were themselves producers of nodule metals from conventional mining and which foresaw their own economic development being threatened by an alternative source of these metals. But whatever the detailed perceptions of participating countries, and however those perceptions may have changed, UNCLOS III has remained fundamentally concerned with achieving international agreement on the regulation of ocean mining. This would encompass the establishment of an international authority to licence and regulate deep ocean mining. Its function would be to select firms for participation in mining operations, to oversee the allocation of sites, to encourage the transfer of the technologies to developing nations, to set production limits on the nodule metals in order to regulate the impact of the new supply on world markets and, finally, to levy a royalty on the proceeds of ocean mining. This authority might even possess an operational wing which would collaborate in nodule exploitation.

Through such an agreement, it was intended to permit mining to proceed within an approved international legal and administrative framework while also offering some control over the pace at which the industry expanded, thus safeguarding the interests of both the consortia and the developing nations. Particularly in respect of this last point, the UNCLOS III proposals would offer incentives to the consortia to undertake joint projects with developing states. Negotiations continued into the early 1980s, and although agreement proved elusive, the industrialized nations found it politically expedient to continue to seek international legislation. But behind the diplomatic activity there was the threat that the United States (followed by others, no doubt) would enact domestic legislation to provide the framework for American firms to begin ocean mining in the absence of international regulation. Such legislation was eventually passed in America. This may have represented a serious indication that the

United States was prepared to proceed unilaterally with deep ocean mining. Or it may have been a tactic to pressurize the Group of 77 to compromise. The internal working of UNCLOS III is still unclear, but the uncertainty generated discouraged the consortia from proceeding rapidly to commercialization, even if technical and commercial factors had been favourable. The consortia would obviously prefer to know where they stood—whether in relation to international or domestic legislation, they wanted a clearly defined framework before committing further resources.

COMMENTARY

The history so far of commercial interest in the recovery and processing of manganese nodules serves as a warning against the danger of being side-tracked when evaluating new marine technologies. At first sight the novelty of the resource and the challenges of recovering it from a hostile environment may seem to constitute the major challenges. Yet, although these specific technical demands will have to be proved on a commercial scale, they are unlikely to be the most critical factors influencing the pace of development. The future of world prices and the establishment of a legal regime to regulate ocean mining seem far more likely to determine consortia decisions on the timing of commercial mining operations. The technological problems may be challenging, the projects may be spectacularly bold and expensive; they may even appear glamorous! But this must not dominate or obscure the commercial and political aspects. It is very much the case with nodule mining. It must not be ignored in other marine technology proposals.

The probable timetable of mining operations, for reasons already stated, is currently unclear. Even on land, large-scale mining operations have long lead times before full production is achieved. For a deep sea mining operation a time lag of ten to fifteen years, following the decision to proceed, would not be unreasonable before the metals were entering world markets. Furthermore new mines are unlikely, even then, to come on stream at more than one or perhaps two per year. This all indicates that even if decisions were taken before the end of this decade it would be well into the next century before metals originating from nodules became a significant proportion of world supply. That is not to reject the possibility of individual commercially viable mining operations within twenty-five years, however.

NOTES

1. The profile of currents and water masses is different in the Pacific and Atlantic Oceans, and this results in the carbonate compensation depth being about

1,000 m. deeper in the latter. In consequence, only some 2 million km.² of the
Atlantic floor is below the compensation depth, compared to 18 million km.²
of the Pacific sea-bed. Hence, manganese nodules are probably more abundant
in the Pacific Ocean.

2. Of course, in a later generation of nodule mining systems, the nodules may
   already have been crushed in the dredgehead in order to improve the efficiency
   of the lift. This would have benefits in handling the materials on the surface.

3. Currently, the supply of nickel, manganese and cobalt is predominantly from
   Eastern bloc and Third World nations. The United States and the Western
   industrialized nations may feel there is more security of supply from ocean
   mining for these metals, but it is unlikely that this concern would be sufficient
   to sustain nodule mining if it proved more than marginally more expensive than
   conventional resources.

4. Surprisingly, the environmental issues associated with the actual deep ocean
   mining operations have not raised as much interest as might have been
   anticipated. This no doubt reflects the fact that the mining is in the deep oceans
   and far from land, and as such its impact will be confined to the deep sea
   ecosystems directly disturbed by the mining. (The debate on Continental Shelf
   mining is discussed in Chapter 3.)

3 Phosphorites and muds

INTRODUCTION

The previous chapter dealt with a large-scale, expensive, deep sea engineering project[1] for the recovery of mineral resources from the seas. But metals and minerals can be, and are being, won from the oceans in other, less grand, but nevertheless effective, programmes. This chapter will look at two of these. One has run for several years with some commercial and great political success and the other, which although still under investigation, provides a contrast to the scale, cost and complexity of manganese nodule recovery. The former programme is the extraction of silver, copper and zinc from muds on the floor of the Red Sea. The latter is the proposal to recover phosphorite nodules from the Continental Shelf and other shallow waters.

PHOSPHORITES

Phosphorous, in many of its compound forms, is an important fertilizer for agricultural use. Agriculture is, indeed, the primary consumer of the world phosphorus supply and phosphorus is not recoverable in recycling processes, nor does a ready substitute exist. Marine sources of phosphorous have therefore attracted much interest; especially since the estimates of those marine resources are enormous! One estimate suggests that economically recoverable reserves are in the order of 30 billion tonnes, which is probably enough for a couple of thousand years at least.[2]

Marine deposits of phosphorus compounds are known as 'phosphorites'. This is a general term that includes a variety of phosphorus minerals differing in both chemical composition and physical form. On land, phosphorus minerals are described as 'phosphate rock' and are found, characteristically, as large bedded deposits; but in the oceans phosphorites are found in four distinct physical forms. These are:

(a) *Phosphorite nodules*. This is the most common sea-bed form. They are often structureless and irregular in shape, varying from small pebbles to slabs up to one metre in length.

(b) *Phosphatic pellets and sands*. These have small grains, up to about one millimetre in size. They contain quite high proportions of clay and mineral impurities. If the carbonaceous content is high these deposits are considerably darker in colour than the nodule form. They are also fairly common.

(c) *Phosphatic muds*. These are similar to the pellets and sands, but are of much finer grain size, thus producing a mud rather than a granular deposit.

(d) *Consolidated phosphate beds*. These extensive sedimentary deposits more or less correspond to the phosphate rocks mined on land.

All these forms are of marine origin.[3]

Phosphorite deposits are distributed widely in the oceans, but they are most abundant on continental shelves and on the continental slope to depths of around 300 m. In the deep oceans they are generally found in depths of less than 2000 m.; but up to 4000 m. is not unknown. The most extensive deposits appear to be off the western margins of the continental land masses (see Figure 3.1).

The precise mechanism by which marine phosphorites are deposited is a matter of disagreement between geologists. Theories of the formation of sedimentary phosphates in seawater need to explain how the mineral could be concentrated sufficiently to produce the extensive deposits which have been found. It is the concentration mechanism over which there is dispute; several mechanisms, both biological and physical, have been proposed, but in the present context the debate is unimportant. There is, however, more agreement as to the general conditions that favour deposition. Water depths of not more than a few hundred metres in low latitude oceans,

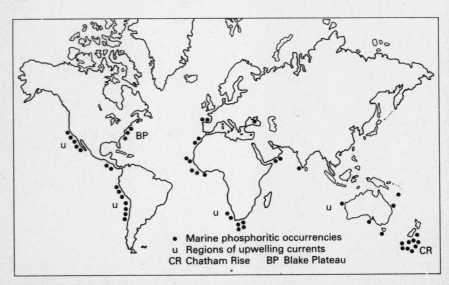

Figure 3.1 World-wide distribution of sea-floor phosphorites. *Source*: M.R.P.

combined with low levels of sedimentation from land, seem to be typical. However, it is in deep ocean water that concentrations of phosphorous minerals are found and these are some thirty times greater than those of surface water. Therefore, phosphorite deposition seems to be associated with up-welling from the deep oceans into the shallower waters, where a suitably undisturbed site will allow for deposition.

By whatever precise mechanism it occurs, phosphorous in various mineral forms is deposited in very large quantities across immense stretches of the ocean-floor and the process has been shown to have been taking place over a very long geological time-scale. Hence, even if the conditions for deposition are very specific, time alone has assured that deposits are now vast, and subsequent tectonic activity has placed them in a variety of locations in the oceans and on land.[4] Some of these marine deposits may be suitable for mining and, compared to manganese nodule mining, this is very simple, requiring little beyond dredging techniques. With deposits available just off shore there is no necessity to venture into deep water and, although the value of the product does not justify the development of sophisticated lifting techniques, there are several choices for shallow-water dredging.

The first option is the 'clam-shell' dredging method which consists of a grab dredge suspended on wires from a surface vessel. The clam-shell makes discrete grabs into the sea-floor. It is a technique applied to sand and gravel dredging and requires a loose granular material on which to operate. But it also needs a thick deposit into which it can take vertical bites. This would not be typical of phosphorite deposits. Nodular or granular phosphorites are more likely to be found as a thin surface coating on the sea-bed. The clam-shell is therefore unlikely to be appropriate for phosphorite recovery.

The second option includes a variety of systems where a series of buckets are dragged across the sea-floor. This may take the form of buckets on a belt attached to a rigid arm (like a conveyor belt), around which the buckets travel down to the sea-bed and then up again to be emptied into the surface vessel (see Figure 3.2). The disadvantage with this is that the weight of the rigid arm restricts its operation to less than 60 m. of water. A more practical arrangement involves buckets attached to a dragline which is dredged across the sea-bed[5] before being lifted back to the surface. This could be achieved with either one or two surface vessels. It would be a cheap and flexible method of operation, except that it cannot guarantee an efficient recovery rate. As the product is of low value it will be important to achieve very high rates of recovery in order for the project to be economically viable.

The third family of recovery techniques are the pumped systems. These can be air-lifts, where the buoyancy of air pumped down from the surface

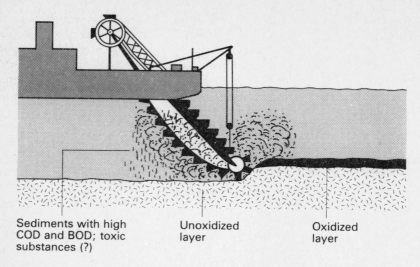

Sediments with high   Unoxidized   Oxidized
COD and BOD; toxic   layer   layer
substances (?)

Figure 3.2 Phosphorite dredging by means of buckets. *Source*: M.R.P.

raises the material from the sea-floor; or they can be direct vacuum suction from surface pumps. Suction dredging is commonly used for clearing harbour approaches in water up to 40 m. deep; with the addition of submerged booster pumps, the depth of operation could be easily increased. For phosphorite nodules this technology appears to be ideal since it provides efficient clearing of the selected site and since it will also lift even the smallest particles. A simple suction pumping system will be adequate; the complex lifting arrangements proposed for deep sea manganese nodule recovery are quite inappropriate for the low-value product and shallow water location of phosphorites.

Once phosphorites have been lifted from the sea-floor, their processing can be very cheap and straightforward. No more need be done than to crush the nodules and form the powder into pellets suitable for direct application to agricultural land as fertilizer. In this unrefined form they have the added advantage of slow-release properties which would reduce the frequency of application. Alternatively, after grinding, floatation and filtration, a phosphate concentrate could be produced which would have a higher unit value and would therefore be suitable for transportation or for further processing into other more refined phosphate products. For these initial, simple processing stages, whether producing pelletized fertilizer or phosphate concentrate, several programmes are considering the feasibility of mobile process plant. Transported by road on several very large trailers, it would follow closely the site of mining operations and so save on transport costs which might otherwise prove a burden if mining began to move away from a permanent processing plant.

Several countries are actively investigating phosphorite nodule mining. Off the west coast of Mexico and Central America, in the Baja California, there are some very large deposits in the form of nodules and sands. These extensive deposits lie in less than 70 m. of water and apparently recovery would be practical. During the mid-1970s, Global Marine undertook an evaluation of the economic prospects from which it was concluded that, in comparison to the 1975 price of land-based phosphate sources, the marine deposits would yield a very respectable profit margin. The Mexican government was particularly interested in the nodules (rather than the sands), which it felt would be compatible with its current land operations. Yet, despite a general belief that Baja California deposits would be the first to be exploited, at prices competitive with land resources, initial exploration was not followed by commercial operations. Official statements by various parties still maintain that preliminary work continues.

In place of Baja California, it was off the coast of southern California that the first serious attempt at commercial mining took place. A subsidiary of the Union Oil Company planned to initiate mining in order to obtain raw material for its processing plants at costs lower than those of phosphate rock transported from mines in Wyoming. Initial costing, on a projected 500,000 tonnes per annum production, appeared to be very attractive. Mining was begun, using a clam-shell grab in 200 m. of water, and large quantities of phosphorites were lifted. However, the project was in fact terminated. The firm claimed that unexploded naval shells had been encountered and that this rendered further mining impossible. It was also rumoured, however, that the yield of nodules (both in quantity and quality) was disappointing and that the marine environment and depth of water had proved considerably more difficult than the feasibility studies had anticipated. In the late 1970s, Union Oil confirmed that it had no further plans to continue its involvement in phosphorite mining.

Unlikely as it might seem, the next most likely country to begin substantial mining is New Zealand. In fact, of course, this should not be surprising. New Zealand is an agricultural nation which lacks its own phosphate resources. Imports of phosphate rock are second only to imports of oil in the country's international trade. But the traditional supplies from other islands in the South and West Pacific and in the Indian Ocean are being rapidly exhausted. So, in the 1970s, New Zealand began to assess the potential of the phosphorite deposits in the Chatham Rise, which is an elevated section of ocean-bed running eastwards from South Island towards Chatham Island. Some of the deposits lying on the Rise would be within New Zealand's two hundred-mile economic zone.

Deposits on the Rise were first discovered in 1952 and are predominantly composed of nodules between 1 cm. and 4 cm. diameter, lying in 400 m. of water. In late 1978 a detailed survey was made of the site and, by extrapolation, the survey suggested the reserves to be of the order of 100 million tonnes, which, although small compared to some deposits, represents several decades of supply for New Zealand, and is well worth commercial interest. Subsequently, studies of economic and technical viability have been begun to see if further commitment is merited. Early results are encouraging and the New Zealand authorities seem keen to pursue research.

Chatham Rise phosphorites contain a lower proportion of phosphorous than the currently imported phosphate rocks. But these rocks require processing for use as superphosphate fertilizer, whereas Chatham Rise nodules can merely be crushed and pelletized prior to direct application as a slow-release agricultural fertilizer. This is very favourable to the economics of phosphorites. Moreover, agricultural field trials have proved that direct application of phosphorite pellets compares very well with superphosphates—all of which reinforces the notion that New Zealand's may yet be the first major commercial mining operation for phosphorites.

For New Zealand, as much as for any other country, the overall economics of phosphorite fertilizer must compare favourably with land-based phosphate rock mining, and the land resources of phosphate rock are so enormous and so widely distributed that strategic considerations of supply hardly apply. Although some specific land resources are beginning to be exhausted, in general phosphorites have to compete directly on cost in all but a few very particular cases.[6] The costs related specifically to the mining technology are unlikely to be important, since they will be little beyond conventional dredging practice. Instead, it will be the overall long-run production costs that will prove most influential and these costs will be primarily a reflection of the transport and processing costs of phosphorite mining.

Unfortunately, the wider prospects for phosphorites are not improved by the recent world prices for fertilizer products which have been depressed throughout the last decade, despite various unsuccessful attempts to restrict supply and thus raise prices. In such a climate it is difficult for any new capacity (either marine or land-based) to be opened up, particularly where the initial capital costs of establishing a new mine will be of more significance than the capital costs of an existing mine which will have already been written off. The current state of the world phosphate market seems to militate against widespread phosphorite development in the near future.

There are some complex environmental issues raised by phosphorite mining which may also restrict the opportunities for extensive mining development. Conventional open-cast phosphate rock mining is highly detrimental to the environment. Legislation, as exists in some states of America, relating to open-cast mining and to the requirements for the reclamation of old mined sites might be to the advantage of phosphorites as an alternative source. Unfortunately, the impact of large-scale dredging for phosphorites in shallow water is still unknown. It will be of two types: there will be the immediate effects of dredging and there will be those related to the subsequent disposal of dredged material after processing. Of the former, whatever method of lifting is used, there will be a physical scouring of the sea-floor and clouds of small-particle sediment will be raised. Apart from the immediate destruction of plant and animal habitats on the sea-bed, the cloud of sediment may have many other longer-term effects. For instance, oxygen levels in the water may be reduced, the penetration of light from the surface would certainly be impaired and heavy toxic chemical substances may be raised from the sea-floor into suspension and into activity. These problems are common to all dredging, of course, but for commercial mining the disturbances of the sea-bed could be very extensive over the large area of the mine-site.[7] However, the disposal of dredged materials after processing is equally a problem. Firstly, it will include large amounts of very fine sediment that might take months to settle. Secondly, if it is a fixed processing plant then the discharge will be sent continuously into the same area, so that that area will never have time to recover. At least the mine-site will be moved on and this will leave the mined areas to return, in time, to a stable habitat. The third problem is that even if the disposal of processed material were eventually concluded in a specific area, there would remain a layer of homogeneous and nutrient-deficient processed material lying on the sea-floor, perhaps to a significant depth, which might not be conducive to the establishment of diverse new habitats.

Clearly, these issues are much more important here than they were in the case of manganese nodule mining. This is because manganese nodule mining is carried out in deep water and in mid-ocean, whereas phosphorites will be taken from relatively shallow water—possibly on continental shelves—where the environmental impact will be more obvious. Particularly off the west coast of America, this is very important and might well have a considerable bearing on the future prospects of phosphorite mining in that area.

The current situation is one in which the overall state of the world market for phosphate products does not encourage the establishment of large numbers of marine phosphorite mining operations. But equally, there are no overriding technical or political reasons why mining should not begin almost immediately in one of several locations (such as the Chatham Rise) if a company or a country believes that specific factor justifies its commencement. The extent of land resources of phosphate rock precludes marine phosphorites from ever obtaining a large share of the market, but as a simple and inexpensive operation, phosphorite mining could be pursued in a number of locations within the near future, so long as the particular features of that location and operation are favourable. Despite restricted opportunities for phosphorite mining, it would not be surprising if it became viable and commercially significant long before manganese module mining did so.

RED SEA METALLIFEROUS MUDS

A second example of marine mineral development to contrast with the case of manganese nodules is the proposal to extract metals from hydrothermal muds on the floor of the Red Sea. It was as recent as 1948 that anomalously high temperatures were first recorded in the deepest waters of the Red Sea and it was not until the early 1960s that it was established that high salinity was associated with the high temperatures. Samples showed the brines to be up to 60°C and of around 250 parts per 1,000 salinity at several sites along the middle of the Red Sea.[8] There appeared to be basins (called 'deeps') along the floor of the sea which acted as traps for the hot brines and, below the brines, muds. Surveys and core samples demonstrated that these muds contained a high proportion of metal-bearing minerals. The muds varied from one deep to another, but typically they were between 10 and 30 m. thick and mainly contained layers of sulphide, oxide and silicate minerals. The sulphide layers in particular are rich in copper, zinc and silver, and these are probably viable for commercial extraction from the muds. Therefore, since the mid-1970s, Sudan and Saudi Arabia, the countries bordering on the Red Sea in the area of the deeps, have been sponsoring research into ways of exploiting these resources.

The formation of metal-enriched sediments seems to be consistent with the terms of current plate tectonic theory. The land masses of Africa and the Arabian peninsula are moving apart and the rift runs, more or less, along the middle of the Red Sea. As the rift widens, the thin sea-floor may crack and allow seawater to seep into contact with hot magma below the oceanic crust. Metallic compounds of the magma dissolve in the heated seawater. As the hot brines rise and meet the colder seawater above, some of the dissolved metals are precipitated out. In certain favourable sites conditions can give rise to the formation of brine and mud pools.[9] Not all deeps have the combination of hot brine and mud deposits, but those that do tend to be found in the deepest depressions of the Red Sea rift between latitudes 19°N and 23.5°N, which places them between the coasts of the Sudan and Saudi Arabia.[10]

The largest, best-known and most thoroughly investigated of the deeps is the Atlantis II Deep, so named after the survey vessel which discovered it. Atlantis II, despite being the largest mud/brine deep, actually only has a surface area of about 60 sq. km. This well illustrates the relatively small and localized scope of any commercial mining of Red Sea muds, since it is probable that if mining were begun it would only be at the Atlantis II site.

The Atlantis II deep provides a good example of the structure of the mud layers. In Atlantis II there are five distinct sedimentary bands within the hydrothermal deposit; these layers reflect the fact that conditions have changed both during the time the muds were being formed[11] and across the area of the basin—different conditions producing different types and distributions of mud. The uppermost layer is an amorphous zone of oxides, hydrosilicates and sulphides with a very high salinity and an average thickness of 4 m. The next layer is of similar thickness with hydrosilicates and sulphides, but the deeper sulphides in this band have a high zinc content. The next zone varies in thickness from 1 to 11 m. and contains iron and manganese-rich oxides[12] and more hydrosilicates. There is then another hydrosilicate and sulphide layer 2.5 to 4 m. in depth and finally the deepest layer—3 m. thick—contains a mixture of oxides and assorted detritus, which, although low in zinc, does have a high copper content.[13] The overall metal content of the muds varies between the layers and between different areas within the deep. Typically, however, metals constitute 10 to 13 per cent of the dry salt-free material. Iron is the most abundant metal, then zinc at between 2 and 3 per cent, copper at 0.5 per cent, and silver at between 50 and 110 parts per million. In the richest muds zinc may be up to 6 per cent and copper up to 1 per cent of dry salt-free material. Although present only in small amounts, the silver represents a very valuable product.

The governments of Saudi Arabia and Sudan became interested in mining the muds in the early 1970s and, through the aegis of the Saudi–Sudanese Joint Red Sea Commmission, sponsored research began in 1976 with a $20 million contract to the West German firm Preussag AG for the continued exploration of Atlantis II and for the commissioning of mining feasibility studies.[14] These studies have produced a new estimate of the *in situ* metal content of Atlantis II. Current estimates now suggest a range of 1.7 to 2.5 million tonnes of zinc, 0.4 to 0.5 million tonnes of copper and between 4,000 and 9,000 tonnes of silver. By mid-1979 test mining of the muds had become possible. The objective of mining is to retrieve muds, of a consistency described as similar to that of toothpaste, from a depth of 2,000 m. of water. The initial studies advised suction pumping of the muds up through a steel pipestring suspended from the mining vessel. The consistency of the muds means that they do not pump easily, so an important aim of the tests[15] was to determine the proportion of seawater that needed to be added to the mud to permit pumping. The pumping apparatus (see Figure 3.3) consisted of a main radial electric pump, for the mud slurry,

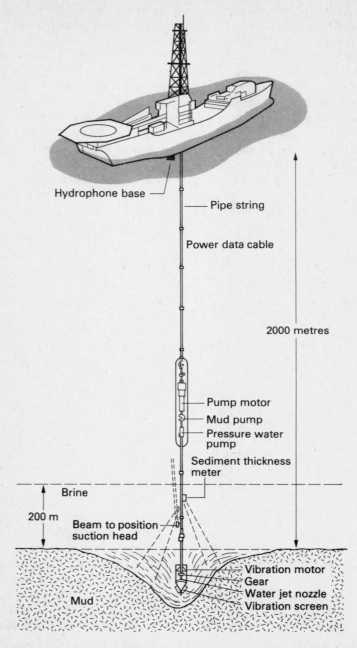

Hydrophone base

Pipe string

Power data cable

2000 metres

Pump motor
Mud pump
Pressure water pump

Sediment thickness meter

Brine

200 m

Beam to position suction head

Vibration motor
Gear
Water jet nozzle
Vibration screen

Mud

Figure 3.3 Suction pumping of muds. *Source*: Georghiou and Ford 1981.

located about 200 m. above the muds in order to be out of the lake of hot, corrosive brine,[16] and a vibrating mining head, with a smaller secondary pump, actually inside the mud itself. The second pump directs jets of the surrounding brines through nozzles into the mud to assist in breaking up the paste and to provide the water for dilution prior to pumping the paste up the pipestring. The pipestring itself for the tests was 13 cm. in diameter and 2,200 m. long (which was the specific requirement of Atlantis II), constructed from steel sections 27 m. in length. The pipe presented few problems. Most aspects of its fabrication, deployment and operation were well understood from techniques developed for offshore oil exploration. Finally, various electronic devices were placed along the pipe to monitor the flow conditions of the slurry within the pipe.

During the trials the technical problems encountered were not severe. They were predominantly electrical and mechanical breakdowns, due to the corrosive nature of the brine and the mud slurry, which is not only hot and salty but also consists of very small mud particles which are both abrasive and clogging. For example, the main pump suffered considerable damage after only two weeks' operation. Notwithstanding these, however, the principle of mining was never seriously questioned. The pumping tests were intended to provide detailed information about parameters of efficient mining practices. These included determining the ideal density of the slurry for pumping, the dynamic behaviour of the pipestring, the positioning of the mining vessel and the impact of heavy sea conditions upon mining, as well as the specific performance of individual components. But these tests were little more than on-site trials of well-understood and well-tried techniques and components. The real novelty concerned the behaviour of the mining head itself within the mud layers. The engineers needed to know the accuracy with which the head could be positioned and controlled within the mud, the depth of mud in which it could operate and the extent of damage caused when the head hit rock rather than mud.

The results of these tests into the operational characteristics of the mining head were very encouraging. Damage occurred, but it was not too serious when the head made contact with rock. The accuracy of mining was good. The head could be dragged behind a slow-moving surface ship and thus cut trenches in the mud. These trenches could be dug close side-by-side at different depths of penetration into the mud. Hence it would be possible to mine preferentially for minerals known to be abundant in a particular layer within the deposit. This implies that excellent control of the mining head is possible, so that precise and efficient mining can be achieved. It was also found that the head could operate deep within the muds and on one occasion it excavated a vertical crater down to rock, 18 m. below the surface of the mud.

Another objective of the trials was to assess the processing of the mud and brine slurry. Early in the preliminary studies it was decided that initial concentration of the slurry would be performed on board the surface vessel. In this case the sheer bulk of the slurry, compared to the relatively small mineral content, renders it uneconomic to transport all the slurry to shore (perhaps 150 km. away) for processing. Some initial processing to concentrate the minerals before transportation was very attractive in this case.[17] This proposal was novel. Concentration would be achieved by froth flotations; this is a common enough mineral engineering process on land, but not at sea. Consequently, the muds raised by the pumping tests were later used to test froth flotation. Flotation was not initially chosen as the mode of concentration since the quantity of solid material in the slurry rendered it impossible to blow air through it, as is the usual practice, to produce the frothing. However, the very fine particle size of the muds and the high salt content ruled out other methods such as gravity separation, centrifuging and leaching. So froth flotation was re-examined and it was decided to use this process, but with the simple expedient of first further diluting the slurry with more seawater.[18]

The apparatus on board the surface vessel homogenized the slurry by mixing until the solid content was uniformly distributed throughout the slurry. This mixture was then pumped into the flotation chambers where it was diluted with seawater and the flotation reagents were added. Once froth flotation was completed, the metal-rich concentrate could be removed from the tanks. These tests proved very successful. The apparatus performed well throughout and with standard flotation reagents the muds were found to have excellent flotation characteristics. The typical concentrate produced had a content of 30 to 34 per cent of zinc, 3.5 to 4.5 per cent of copper and 500 to 700 g. per tonne of silver. (The concentrate also had 14 to 18 per cent of iron, 0.5 per cent of lead and 26 to 28 per cent of sulphur.)

One problem that had been anticipated turned out, in the event, to be no problem at all. It had been feared that the motion of the vessel, especially in heavy seas, would prevent flotation from working efficiently, and plans were laid to mount the whole flotation apparatus on a gimbal to counteract any motion of the ship. In practice, however, the ship's stabilizers were found to be adequate for all but the most severe weather conditions.

The flotation tests provided data by which to cost the processing sector of metalliferous mud mining. In particular it was found that good overall recovery rates could be achieved even with comparatively low-grade muds (i.e. those with an initial zinc content of as little as 3.4 per cent), as well as with the higher-grade ones containing around 5.5 per cent of zinc. This was important since it suggested that more of the muds might be commercially exploited than had previously been anticipated.

The third objective of the mining trials, and the third reason for raising large quantities of mud, was to ascertain the behaviour of the tailings disposed back into the sea after the removal of the metals. Both Sudanese and Saudi governments had been very insistent from the outset that environmental damage would not be tolerated. The impact of mining and processing had to be shown to be slight before commercial mining could proceed.[19] Since the mining operation itself would be restricted to the Atlantis II basin, the specific impact of mining would be restricted to a relatively confined site and the overall damage was considered, therefore, to be negligible within the context of the whole Red Sea. However, more concern was focused on the problem of disposing of the exhausted tailings after processing. There were two specific issues here. Firstly, the surface and near surface marine communities were to be disturbed as little as possible. Secondly, the coral reefs along the Red Sea coasts must suffer no damage. A further, general condition stipulated that no other habitat or community was to be unduly affected in order to meet the two primary criteria. Together with the environmental concern, there was the practical point that tailings could not be dumped back into the mine-site since this would ultimately degrade the deposit and render mining uneconomic. In effect, these criteria meant that disposal had to be carried out at a depth great enough to avoid surface communities and great enough to allow the tailings to sink before being swept onto the reefs, yet not so deep as to prevent the tailings from becoming thoroughly diluted and dispersed before reaching the sea-bed. Tests showed that tailings pumped out at around 400 m. down[20] sank rapidly enough whilst achieving wider dispersal than was expected. This was considered to be satisfactory and the overall biological disturbance was acceptable. For commercial operations, however, the high rates of disposal required may be best achieved by pumping the tailings at a greater depth and by the coagulation of the sediment to encourage rapid settlement. Nevertheless, the rate of dispersal of the tailings cloud (which is actually enhanced by the high salinity of the tailings) would still be sufficient to avoid serious localized damage.

In summary, the various trials carried out demonstrated that muds could be raised and a metal-rich concentrate produced at sea. Following transportation to the shore, the concentrate still requires to be refined. Initially, smelting processes were considered, but it became clear that pyrometallurgical routes required considerable reduction of the concentrate's salt content. In neither Sudan nor Saudi Arabia could huge volumes of fresh water be spared merely to wash the concentrate. So attention was turned towards the relatively new techniques of metal chloride leaching. Development had been carried out in the United States and mineral concentrates of composition similar to those of the Red Sea muds had been processed. Four different leaching routes were proposed and tests were begun to ascertain the details of process design, capital costs and operating costs for a commercial plant. All four processes performed well and achieved similar yields of around 99 per cent zinc, 95 per cent copper and 90 per cent silver recovery from the concentrate. These yields were well beyond what could be achieved with alternative operations. The final choice between the four individual leaching processes is a matter of detailed assessment and adaptation to the specific requirements of both the Red Sea mud and the countries concerned.

The commercial future of Red Sea mud mining appears to be very sensitive to the efficiency of metal recovery and the market price of silver, zinc and copper. The absolute value of the metals and the magnitude of price fluctuations in world markets mean that yields and prices, rather than the costs of research and development or capital and operating costs, will determine the return on investment for mining. Froth flotation and chloride leaching offer the high yields, but metal prices are less predictable, especially in the short term. However, in the longer term, overall trends are usually steadier. Zinc is an important industrial metal which, contrary to some fears, has not lost its traditional markets to substitute materials, and future demand should at least remain stable and might perhaps show small, steady growth. Copper's excellent conductivity for both electricity and heat make it an important and not easily replaced metal for many industrial uses. Some predictions suggest continued rapid growth in demand into the next century. Existing reserves would be hard pressed to meet such demand and although rising prices may call other new sources of copper into production, the prospects for copper derived from the Red Sea remain very encouraging. However, of the three metals, it is silver that is assured of a high constant demand. Current world demand for silver is growing, especially the demand from the photographic and electronic industries, and that demand is already exceeding annual production, so new supplies will find a ready market.[21]

For all three metals, demand appears to be steady and, notwithstanding short-term fluctuations, the overall price of the metals on the world markets ought to remain high enough to produce an acceptable rate of return for investment. But, more clearly perhaps than in any other new marine project, narrow commercial criteria cannot be sufficient to understand the aims and prospects of the programme. Whereas some marine projects in the West have been vaguely justified in terms of the strategic need to ensure supply or to control world prices, Red Sea mining has a much clearer economic, political and even religious purpose that transcends immediate commercialism.

In the late 1960s Saudi Arabia somewhat anticipated subsequent issues at the United Nations Conference on the Law of the Sea by enacting legislation that not only reaffirmed Saudi rights to the continental shelf around its coast, but also extended Saudi claims to rights over the Red Sea sea-bed beyond the continental shelf. In effect Saudi Arabia had unilaterally appropriated the resources of the Red Sea. But this was a tactical ploy, for the Saudis immediately made it clear that they would welcome collaboration with neighbouring states in the exploration and exploitation of Red Sea mineral resources. Consequent upon this legislation and the corresponding desire for co-operation, Saudi Arabia and Sudan signed an agreement in May 1974 that granted each country sole rights over the sea-bed from its coast to a depth of one thousand metres. The area between the two exclusive zones, and any resources lying therein, would be the common property of the two countries (see Figure 3.4). At a depth of two thousand metres, the brine pools and associated basins of metalliferous muds are well within the common zone. To facilitate the speedy development of any commercial resources of the common zone, the Saudi–Sudanese Joint Red Sea Commission was established and its immediate task was to handle the question of the muds.

The Saudi government provided funds for both the running of the commission and the initial development programme. This was an amicable agreement for, if a line were drawn down the geographical centre of the Red Sea, most of the brines would actually lie in the Sudanese half. The Sudan, however, would not have been able to finance exploration. The bilateral agreement suited both parties, but it is more than a mere commercial expedient. Collaboration was explicitly intended to foster and cement Arab and Islamic solidarity by the demonstration of co-operation and by its example as one of the ways that Arab states could seek to reduce their dependence on oil revenues. Had the commercial outlook for mining been less encouraging, the Saudis, in particular, might still have been keen to proceed in order to achieve these wider aims.

During 1986/7 it is intended to run a full pilot mining operation producing concentrate to feed a land-based processing plant. These trials will aim to prove the technical feasibility of larger-scale mining, to assess the four individual leaching processes on a scale more comparable with commercial operations, and to confirm that the environmental impact is indeed minimal. If all these tests are resolved positively, commercial mining could begin by the end of the decade.

Although the daily production of commercial mining will be of the order of 700 tonnes of dry salt-free concentrate (this is some seventy times more than even the pilot plant will produce) from the single mineship operating in Atlantis II, this still implies that considerable scaling-up is needed. Neither the surface vessel with its on-board concentration plant nor the pipestring (which will be around 150 cm. in diameter) have been constructed and deployed. But the engineering requirements of full commercial mining are not believed to be insurmountable. Both political will and finance are available and the potential gains, in terms of new skills

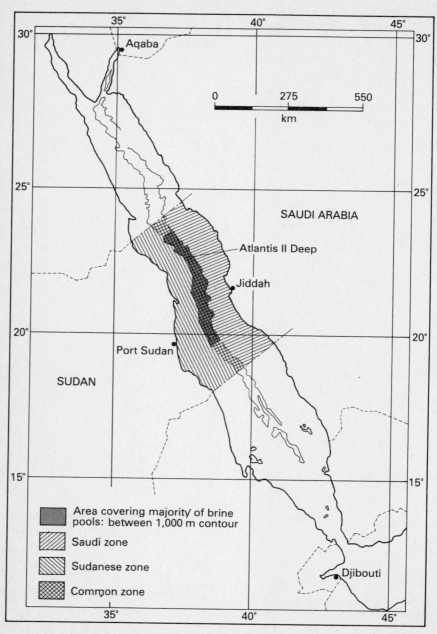

Figure 3.4 Red Sea Commission area. (Adapted from Simnett, 1982).

required by the work-force, industrial diversification, Arab unity, as well as the manufacture of a valuable commercial product, are sufficient to suggest that every effort will be made to ensure the programme is brought to a successful conclusion.

COMMENTARY

The examples of Red Sea metalliferous muds and marine phosphorites both contrast, in their own ways, with the case of manganese nodule mining. Phosphorites are a low-value product. They can be scooped off the sea-floor with fairly conventional dredging apparatus and then used directly as agricultural fertilizer. Red Sea muds contain a more valuable material and the exploitation of that material is less simple than for phosphorites, yet bilateral co-operation on the questions of sea-bed mineral rights and collaborative exploration have resulted in rapid progress towards commercial mining in little over ten years. But, even if Red Sea mud mining is technically more sophisticated than phosphorite mining, it is envisaged as operating on a small scale compared to phosphorite mining and both of them are tiny programmes in comparison to the proposals for manganese nodule mining. It is, however, the relative simplicity and the relatively small scale of phosphorite and Red Sea projects that means that either of them could be delivering a product onto world markets and by the end of the century. It is unlikely that any of the more exotic marine technology proposals, with their grand and expensive programmes, will be operational before that. This illustrates an important point. It is misleading to concentrate on the expensive, high-technology projects which have thus far monopolized so much of political and media attention.

NOTES

1. Chapter 5 also deals with another large-scale programme. It too is not necessarily typical of all the opportunities for marine technology.
2. The Royal Institute for International Affairs in 1982 estimated the identified land resources to be 130 million tonnes. The scale of the marine reserves is clearly considerable.
3. It is also possible to find phosphorites composed of guano deposits which originated above sea-level but which have subsequently been submerged.
4. The range of ages of deposits may partly account for the range of physical forms in which marine phosphorites are found.
5. These systems are less sophisticated versions of the 'continuous line bucket' mining option proposed in connection with manganese nodule lifting; see Chapter 2.
6. This is the situation that applies to New Zealand's traditional supplies and so encourages that country to look seriously at phosphorites.

7. Surprisingly little is known about the long-term consequences of dredging. There is some evidence that, apart from the initial destruction, dredging can be beneficial to the overall ecosystem. Such benefits may include the re-oxygenation of sediments after resettlement, the release of nutrients into suspension which had been trapped in the sea-bed, and an improved oxygen distribution in the water column owing to mixing.

8. By the early 1980s, eighteen such hot brine 'deeps' had been surveyed.

9. The brine pools have been likened to lakes of hot, dense and very salty water lying in natural basins in the sea-floor below the ordinary cold, deep seawater. In the Red Sea some of these deeps also trap hydrothermal muds.

10. This convenient situation, in which only the two countries are involved, has permitted rapid bilateral political agreement on a programme of research into the prospects of mining the muds. More will be said of this below.

11. Estimates suggest that the muds of Atlantis II are the product of about twenty-five thousand years of activity. The process may now only be continuing at a few sites within the basin. .

12. There is no interest in extracting these metals commercially.

13. The internal structure of Atlantis II is the most complex of any of the deeps. The muds show a variety of physical as well as chemical forms and even vary in colour. White, orange, red, grey, green, black and beige muds have all been found.

14. Oceanographic studies of Atlantis II and the other deeps continued during this period, but commercial interest was confined to Atlantis II.

15. A subsidiary, but major objective, was to lift several tens of thousands of tonnes of mud for processing tests and tailing discharge tests. See below.

16. At this depth the seawater temperature was down to 22°C, as compared to 60°C in the deepest brines.

17. This is in contrast to phosphorite nodules where processing at sea was not considered to be necessary since, although the product was of comparatively low value, it was, in effect, already concentrated by efficient mining practices.

18. The additional dilution, including the dilution caused by the flotation chemicals, reduced the solid content of the slurry from 50 g. cm.^{-3} to 25 g. cm.^{-3}. Eventually it was realized that froth flotation was the only concentrating process that was not adversely affected by the high salinity of the slurry.

19. The extent to which the Saudi and Sudanese authorities acknowledged joint responsibility for the protection of the Red Sea marine environment is one of the features of political concensus that has characterized this programme. More generally, in the mid-1970s, other nations around the Red Sea agreed to be jointly responsible for monitoring any environmental disturbances in the area.

20. The current and water mass profile of the Red Sea is very highly stratified, with little mixing between upper and lower waters; hence, nutrients are not easily transferred into the lower levels. This results in a lower level of biomass in the Red Sea than at comparable depths in other oceans. Thus, tailings can be disposed of at relatively shallow depths in the Red Sea with little environ-

mental impact. Such shallow water disposal is not appropriate for other deep ocean mining projects outside the Red Sea.

21. The scale of mining envisaged would only increase world supply of any of the metals by about one per cent. So there is no danger of world prices being depressed by the additional supply.

4 Other minerals from the sea

INTRODUCTION

In the two preceding chapters we have considered three major prospects for extracting minerals from the sea-floor. These programmes were discussed in some detail in order to contrast the high-technology requirements of manganese nodule mining with the much simpler, smaller-scale engineering of the other two and in order to emphasize the crucial importance of the political and legal framework of ocean mineral developments. But manganese nodules, Red Sea muds and phosphorites are far from being the only options for ocean mining in the future. Although these three have so far received the most attention, there are a number of other schemes that may in time prove to be of greater importance to marine technology and mineral supply. In this chapter a brief review will be made of five other schemes, not all of them by any means primarily involving complex technology or solely concerned with high-value metals. Indeed, one scheme involves little more than the possible extension of dredging for sand, gravel and other bulk building materials.

All three examples used in Chapters 2 and 3 were of resources which lay on the sea-floor and for which some more or less sophisticated dredging and lifting technology was required. Two of the projects described below are of that type. Sands and gravels are loosely-packed deposits on the sea-bed and radiolarian oozes, which are organic muds that may provide a source of ceramic material, are similarly found as a sea-floor deposit. Technically, the retrieval of either only requires the development of the appropriate lifting system. However, there is a further mineral deposit which is located below the sea-floor, the mining of which will obviously necessitate more than merely lifting the mineral off the ocean-floor. These hydrothermal sulphide deposits have been discovered in the Pacific and contain high proportions of industrially valuable metals, but since they are below the sea-floor and in the form of a conventional mineral ore, rather than being either nodules or muds, there are severe difficulties to be faced in mining the deposit even if the necessary lifting technology is available.

Solid materials on or under the sea-bed do not exhaust the oceans' mineral resources. There is an enormous range of metals and minerals dissolved in seawater and some of these could be worth extracting commercially. One clear advantage of this is that, with the exception of minor local variations, the minerals in seawater are evenly distributed throughout the body of water. Mineral extraction can therefore be undertaken from coastal installations; there is no need to lift impure ores from the sea-bed several kilometres below the surface of the oceans. Unfortunately, the concentration of most minerals in seawater is so low that extracting them is difficult and very expensive. Although salt can be precipitated out of seawater simply by evaporating the water, other minerals will require more sophisticated and often energy-intensive processes. In this chapter we will deal with one case where a simple, primarily evaporative, technology has been used for the commercial extraction of potash from the Dead Sea. But this is only possible because of the abnormally high mineral concentration of the Dead Sea waters and other technologies are required if minerals of lower concentrations are to be extracted from seawater. Electrolysis is an option which has been used for many years to extract some elements, most notably bromine, from seawater, but it is expensive and unlikely to be suitable for a general extractive marine technology. So, at the end of the chapter attention will be given to a new proposal to remove metals from seawater by the action of ion exchange resins which are polymers that can selectively remove a specific metal ion from solution. Ion exchange is a long-term option, but if it were developed for commercial processes it might open the way to the extraction of almost any element from seawater and so render largely redundant the need for deep ocean mining projects in the long term.

However, before considering the exotic options for the next century we will return to the more prosaic dredgings and lifting technologies of the nearer future.

SANDS AND GRAVELS

Dredging bulk materials in shallow water is hardly a novel operation, and the technology required is relatively simple. With the advent of new lifting systems, developed for other materials in deeper water, dredging for low-value bulk building materials, such as sand and gravel, might be extended. Britain is one of the countries that has dredged for these materials off its coast for many years, and with an expansion of marine engineering in the North Sea it would very probably seek to increase these activities. However, the low value of sands and gravels requires that the cost of recovery be kept low, otherwise production could easily become uneconomic. Even if a lifting technology for deeper water sites were available and cheap to operate, the increased transport costs of working farther from shore could render it unprofitable. Indeed, currently it is unusual for dredged sands or gravels to be used more than a few kilometres inland. Whether the burden of cost increases falls to lifting or to transportation, it is unlikely that the value of dredged bulk materials will be sufficient to justify the development of a lifting technology to operate in waters deeper than the present limit of about 50 m. Although the reserves of sand and gravel on land are very large, there might, in the future, be increasing pressure, both economic and environmental, against the use of land for quarrying. Such circumstances would probably increase the market for marine sand and gravel, but one would not expect it to be enough to support continued development of deeper water dredging technology.

The primary reason why marine sand and gravel may become increasingly important is related to the future of another new marine technology and is not particularly dependent upon traditional building uses on land. If the construction of 'artificial islands'[1] in areas such as the North Sea becomes significant and widespread in the early decades of the next century, then sand and gravel dredged from near the construction site will be the main source of bulk material for filling the artificial islands. This would consideraby increase the demand for these materials and although it would not necessarily involve lifting from beneath deeper water, it would surely act to stimulate a review of recovery techniques in all conditions. It may even lead to the adoption of more sophisticated technologies from other marine enterprises.

Since Britain is already dredging extensively and since the North Sea may become an important area of artificial island development, British coasts and waters may suffer the consequences of increased dredging activities. There is no doubt that dredging for sand and gravel is the marine equivalent of open-cast mining. It is very destructive of the marine environment, but the value of the product precludes the employment of more selective technologies. Dredging scours large tracts of sea-bed and lifts into suspension vast clouds of fine sedimentary particles. In the relatively shallow waters over continental shelves this disturbance can have serious consequences at all depths of water, not only on or near the sea-floor itself. The immediate damage to the habitat is severe and recovery is estimated to take two or three years,[2] but of more concern would be the impact of the sediment cloud itself. Drifting in suspension, it would adversely affect any fishing grounds into which it penetrated, and even short-term disruption of fishing could have disastrous consequences for fishing industries already struggling with quotas and depleted stocks.

The low value for volume of dredged bulk materials tends to exclude the development of new lifting technologies specifically for those materials. However, the expansion of construction technologies for use at sea, such as those for artificial islands, may encourage the adoption of lifting systems developed initially for other mineral resources. Nevertheless, even if the economic and technical aspects of extending dredging can be overcome it may be that environmental considerations will ultimately limit the scope for expansion.

RADIOLARIAN OOZES

Radiolarian oozes are deep-water minerals which lie on the sea-bed and are therefore retrievable by dredging or suction methods. 'Ooze' is the term applied to fine organic sediments on the sea-floor which are composed of either calcareous or siliceous minerals from the skeletal remains of small marine animals, mixed with deep-sea muds and clay. Radiolarian oozes specifically are siliceous sediments with a high silica content. When considered as a mineral the deposit is called radiolarite and extensive deposits have been discovered on the floor of the Pacific Ocean in association with manganese nodules.

In the mid-1970s the Scripps Institute of Oceanography in California began to take an interest in radiolarite as a possible source of new ceramic materials. The radiolarian skeletons can be bonded together to give a light-weight ceramic which exhibits the usual ceramic properties of chemical and heat resistance. The special advantages of radiolarite ceramic are that it can be bonded at lower temperatures than most conventional ceramics, thereby saving energy costs in production, and its strength-to-weight ratio for compressive forces is very good, comparable indeed with cellular concrete. These characteristics render it an attractive material for building purposes as an aggregate in concrete where lightness and strength are needed. There is also, of course, its potential utilization as a traditional ceramic material where it may be desirable to combine the resistant properties of a ceramic with the lightness and strength offered by radiolarite ceramics.

Since radiolarian oozes have been found on the sites of manganese nodule deposits it has been proposed that oozes could be mined together with the nodules. The argument is that radiolarite could be lifted out for little increase in the overall costs of mining and that it could subsequently be processed and transported within the framework of an existing nodule mining operation. This possibility notwithstanding, the problem remains that radiolarite is a low-value material and, even if retrieval could be under-taken as an inexpensive adjunct to nodule mining, the cost of transport from mine-sites in mid-ocean to land might well be prohibitive. Certainly one could not anticipate the market for radiolarite being able to expand far beyond the margins of the Pacific Ocean.

At present radiolarian oozes have no future. The manganese nodule programmes are dormant and it is impossible to foresee any viability in lifting the oozes by themselves, or perhaps more pertinently, if it were possible to lift the oozes then the nodule programme would not be dormant! There is no doubt that the future commercial exploitation of this ceramic mineral resource is only feasible as part of another, independently viable, operation.

HYDROTHERMAL SULPHIDES IN THE PACIFIC OCEAN

Extensive deposits of polymetallic sulphide minerals were discovered as recently as 1977 by a team of American scientists along the Galapagos Rise in the eastern Pacific Ocean (see Figure 4.1). The deposits were found associated with hot water vents on the sea-bed. Since then hydrothermal sulphides have been identified at other locations along the East Pacific Rise and within the Guaymas Basin off San Diego. The origin of the minerals appears to be related to the location of the deposits, which correspond to the axis of the spreading centres of the ocean-floor. At these spreading centres new oceanic crust is formed by the welling up of basaltic magma between adjacent tectonic plates and, as the new sea-floor spreads, the heat and the forces generated cause fracturing of the rocks. Seawater can percolate down through the fractures, where it is heated to temperatures great enough to leach metals out of the rocks. The metal-enriched brine is then forced back through the vents in the sea-bed at temperatures of up to 350 °C. On meeting cold seawater again, the hot metal solution is cooled and the metals are precipitated out as metallic sulphides which collect as vast deposits inside and outside the vent, often forming tall chimneys around the mouth of the vent itself.

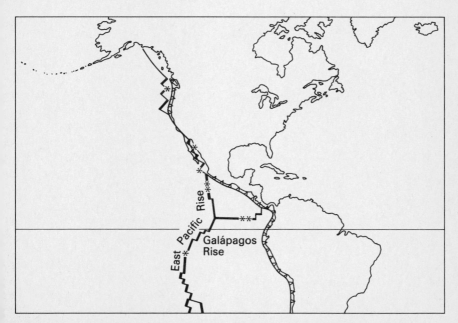

Figure 4.1 Sites of hydrothermal and polymetallic sulphide deposits. *Source*: Ford, Georghill and Cameron 1983.

The sulphides are found at depths of several kilometres and the most thoroughly investigated site is the Galapagos Rift Valley, at a depth of two kilometres. Surveys reveal that the metal contents of the deposits vary considerably, but in the Galapagos region the averages are 38 per cent iron, 6.5 per cent manganese and 0.3 per cent aluminium, together with other metals, often including gold and silver, in smaller quantities. From some sites on the East Pacific Rise, however, samples have contained over 40 per cent zinc. The proportion and distribution of metals is related to the water temperature from which precipitation has taken place. For example, at around 350°C copper and zinc are deposited, but at lower temperatures (between 270°C and 300°C) zinc predominates since the copper will already have been precipitated much deeper below the sea-bed where the water was hotter. By virtue of this process many of the largest deposits are located below the sea-bed. If the temperature at which precipitation occurs is reached while the water is still under the sea-floor, then that is where the deposits will be found. These deposits can be extensive and are not subject to erosion as are the chimneys and other deposits on and above the sea-bed. Therefore the largest ore bodies will most probably be found where the water issuing from the vents is at less than 270°C, since if the water has cooled to that temperature then precipitation has already occurred at some depth below the vent.

In consequence of their deposition below the sea-bed, the retrieval of polymetallic sulphides of hydrothermal origin presents severe problems. For manganese nodules a dredging/suction method was feasible (even if not actually demonstrated to be commercially viable) because of the physical size and distribution of the nodules on the sea-floor, but sulphide deposits are neither conveniently nodular nor, for the most part, do they lie open on the sea-bed. The sulphide deposits have more in common with a conventional rocky mineral ore than with the muds and nodules of other ocean mineral deposits. Simple dredging and suction lifting is of no use in this case and although a modified version of the lifting technology proposed for manganese nodules might be suitable where the sulphide ore above the level of the sea-floor is already in small pieces, it would not tackle the more important rocky precipitate below the sea-bed at the cooler vents.[3] A genuine deep ocean mining technology is needed to extract the ore and at present no such technology is available. Even accurate surveying of the quantity and quality of the ore is difficult. Core samples have to be taken by drilling from surface vessels and at depths of in excess of 2,000 m. this is expensive and there are few oceanographic survey ships that are available for commercial exploration. Although everything points to a vast and rich mineral resource, much surveying will be needed before the full extent is determined. Only then will the parameters needed for the operation of a mining technology for sulphides be known.

As simple dredging does not appear to be a feasible mode of retrieval, it has been proposed that in the future sulphides might be obtained by tapping directly onto the vent and pumping the hot metal-rich solution straight up to the surface as it emerges from the vent. This does have considerable advantages over the option to mine out of the rock itself. Not least, it is convenient that the metals are already in solution and ideally suited for pumping to the surface rather than having to be lifted as a rock ore. However, apart from the obvious technical problems of handling the metal solution there are uncertainties about the rate at which hot brines emerge from the sea-bed. The current rate of deposition may be slow and the huge deposits may reflect the long time-scale of deposition rather than current high rates of sulphide formation. The flow rates may not be sufficient to sustain commercial exploitation and even if they are this method cannot exploit the deposits already precipitated out as rock.

Even in the absence of a mining system, sulphide deposits have aroused interest as a source of metals. Commercially, it would be silver and zinc that would be the major targets. In part, the interest is a consequence of the loss of momentum in the manganese nodule programmes. The failure to achieve satisfactory international agreements on sea-bed mineral resources outside territorial waters has adversely affected nodule programmes since the consortia have been reluctant to commit the resources required for the final push to commercialization.[4] Hence those consortia, and the firms that comprise them, have a financial and historical commitment to marine mineral technology which is not now being fully utilized. They now have experience of marine mining development and see hydrothermal poly-metallic sulphides as a new venture in which to employ the experience gained during the earlier commitment to manganese nodule mining. In this there is the danger that any interest in sulphides is merely an attempt to justify the continued existence of marine mineral technology departments within a firm's research and development establishment. Alternatively, it might be thought that any chance, however small, of recouping some of the expenditure from the manganese nodule programme is worth pursuing and so sulphide mining research could be given a few years to show that it is not going to be a case of throwing good money after bad. But the commitment might be cosmetic.

On the other hand, there might be a genuine desire by the consortia to use the experience that has already been gained in order to develop serious commercial mining of sulphides, rather than of manganese nodules. This assessment may be near the truth if the consortia believe that political rather than technical or commercial factors were predominant in stalling manganese nodule research. Certainly these are firms with a history of solving problems in high technology operations and for this reason they may not believe the technical problems of mining hydrothermal sulphides to be intractable so long as the political and legal climate is favourable. If that is the case then sulphides carry one considerable advantage. Although the deposits are found in deep oceans, part of the East Pacific Rise runs along the coasts of northern California and Oregon only one hundred kilometres offshore and any mineral deposits on it are clearly within the economic zone of the United States.[5] Mining could take place within the context of American domestic legislation with no reference to wider international debates over the mineral resources of the oceans. Indeed, sites off Oregon have recently been put out to firms on lease and surveys of the resources have begun.

Japan has shown itself consistently prepared to explore all aspects of marine resources until an option becomes unworthy of further pursuit. In keeping with this philosophy, Japanese government finance has been available for feasibility studies carried out by shipbuilding firms and mining companies in collaboration with the Japan Marine Science and Technology Centre. As yet no deposits have been identified close to Japanese coasts but there are several sites where the current theories of sulphide formation would suggest that minerals could be expected to be found and some of these sites lie within waters over which Japan would claim exclusive mineral rights. Like the United States, Japan is aware of the benefits of avoiding the exploitation of minerals in international waters where at all possible.

It is impossible to anticipate the future of polymetallic sulphide mining this early on. Only Japan and the United States are undertaking exploratory surveys at present.[6] The interest in the United States may be no more than transitory as the consortia seek to employ resources for which they have no other use at present. Nevertheless, it may be that the problems of mining the sulphide will be overcome by a committed research effort and the deposits may come to provide a valuable and high-grade source of metals.

Although no attempt can be made to predict the cost of sulphide mining, the parameters for commercial viability are clear enough. As was discussed in relation to manganese nodules and Red Sea muds, the target metals of silver, copper and zinc all seem to have a future of steadily increasing demand. So sulphide metals may well find a place supplying that increased demand, so long, always, as the production costs are more or less comparable with conventional sources. This is the major imponderable. Doubtless, with sufficient investment, the technical problem of mining from below the sea-bed under 2 km. of water can be solved, but it may only be possible at a cost which prohibits a viable commercial future for many decades to come.

POTASH

The Dead Sea is, of course, not a sea at all, but it is still an important area in the development of truly marine technologies. The Dead Sea experiences very high evaporative rates from its surface and thus, together with the large quantity of mineral salts dissolved in the waters flowing into the Dead Sea, results in the lake having very high concentrations of minerals in solution. The extraction of minerals from the Dead Sea has been in progress for many years, but although the conditions there are not in any way typical of conditions in the open ocean, techniques developed for the Dead Sea may in the future form the basis for extractive technologies for the lower concentrations of minerals in seawater.

The primary target of mineral extraction from the Dead Sea is potassium chloride (potash). It has some industrial applications but it is more important as a fertilizer. The potash is extracted from the Dead Sea by fractional crystallization—a method by which the different solubilities of the salts are utilized to separate the desired product. Since a solution containing dissolved minerals is subject to evaporation, the concentration of the minerals will increase and, depending upon its solubility, each mineral will have a unique concentration beyond which it does not remain in solution. At that concentration the mineral will be precipitated out of the solution. Thus, by controlling the evaporation it is possible to remove a succession of minerals from a solution containing several dissolved salts. By this process it is possible to evaporate the Dead Sea waters until a complex salt called carnallite is precipitated. This salt contains, among other minerals, potassium chloride, and with further chemical treatment this can be separated from the other chlorides in carnallite.

This is not a complicated technology by the standards of chemical engineering and on the shores of the Dead Sea the main input of energy for evaporation is provided directly by the sun. Hence the operation is commercially viable. But the technology would not be suitable for the low concentration of minerals in seawater. The energy input would be far too great and would render the whole process wholly uneconomic. But there are other minerals in the Dead Sea and some of them are salts of the industrially valuable metals such as magnesium, sodium, phosphorous and calcium. In the extraction of the potash much of the basic processing has already been undertaken and it may be feasible to add a more technically sophisticated refining operation to the scheme in order to retrieve these metals. Such a proposal would certainly not be viable were not the concentrations of the metals so high and had not the initial processing been completed under the potash extraction, but from a technology developed for the high concentration of the Dead Sea minerals there might be spin-offs for a marine technology at the lower concentrations of seawater.

The other conventional method which might be employed to extract metals from seawater is an electrolytic one. It has been employed in specific circumstances. But in general both evaporative and electrolytic processes are too energy-intensive for all but the most favourable constellation of factors.[7] The world market prices of other minerals would need to rise dramatically before it became viable for a wider range of products. This may of course prove to be the case, but it cannot be relied upon to ensure the future of commercial mineral extraction from seawater. For this a new approach is needed which will provide substantially reduced production costs and it is unlikely that, however sophisticated the development of conventional evaporative (or electrolytic) methods become, conventional technologies will be able to deliver such large cost reductions. In the section below we will consider one new technology which might lead to the new approach that is sought.

MINERALS FROM SEAWATER BY ION EXCHANGE

Seawater contains a large selection of elements and the most valuable ones, such as gold, have been subject to many schemes for extraction. But as we have seen already, conventional processes are rarely viable owing to the low concentration of the minerals in seawater, and the costs of energy and reagents are prohibitive. Attention has therefore turned to a variety of alternative extraction processes which would operate at low mineral concentrations and one technology, though still in its early development, is attracting considerable interest because it may avoid the cost penalties of traditional processes.[8] This technology is based on 'ion exchange' where resins can be produced which will efficiently and selectively remove the specified metals from solution. The resins will probably be polymers and the exchange is made between an ion in the polymer and the metal ion in the solution. Once the metal ions have been extracted and concentrated in the resin, then more conventional solvent chemistry will usually be sufficient to remove the metal from the resin.

Ion exchange resins may render it feasible to remove low concentration metals from seawater as a viable proposition, but it will not be the old favourite gold that will be sought. Instead, it will be the high technology metals, such as molybdenum, vanadium and, primarily, uranium. Apart from their undoubted industrial value these metals are appropriate targets for commercial extraction since they are present in seawater in higher concentrations than gold is. Extensive research has been undertaken towards the development of an ion exchange resin to extract uranium. In laboratory tests the principle has been demonstrated and there appear to be no fundamental impediments to the use of ion exchange extraction. That appropriate ion exchange resins can be produced to remove uranium from seawater is only part of the task. In commercial operation the resin must be suitable for large-scale and long-term processing.The important research and development aim is now to produce a resin with the properties needed for commercial operation rather than laboratory demonstrations.

In view of the low metal concentration in seawater, a commercial plant would require a very large volume of water flowing through it in order to show a significant yield of metal.[9] These flow rates would prove expensive if achieved by pumping, unless that pumping were underaken anyway for another project. Such large flow rates may be available in conjunction with either an OTEC facility (Ocean Thermal Energy Conversion) or a large desalination plant, but in their absence the use of natural tidal and current flows is an attractive alternative. Sites with high tidal flow rates would also be attractive as locations for tidal power schemes and so the possibility of a joint operation of two marine technologies may make feasible projects which individually remain uneconomic and impractical because of the large scale of marine engineering necessary. Other than at sites of large tidal range, naturally high water flow velocities can be found in straits where water is forced into a narrow channel between land masses, hence raising the speed of flow. This may be preferable to tidal flow since, in a closed basin or estuary, seawater that has already been depleted of its metal contents may be recycled and so reduce the efficiency of the operation, whereas in open sea straits it is less likely that depleted water will re-circulate.

The scale of the project is not inconsiderable: even a large desalination plant by current standards would only have a water flow rate equivalent to the production of one tonne of uranium per year. So it is obvious that a proposal to produce 1,000 tonnes per year involves a huge construction effort even if a suitable site can be found.

The uncertainties of costs this early in the research phase are such that forecasting economic viability of uranium from this source is impossible. Estimates currently vary from five times the cost of land-based sources up to fifty times as much. At the lower end of the range a continued rise in world demand and the desire to secure supplies might make the project attractive. But even strategic imperatives might be insufficient if the product exceeded the cost of land-based resources by an order of magnitude. There is no doubt that ion exchange provides some hope of processing within the lower cost range and this is something beyond the scope of conventional extractive technologies. But the other imponderable is whether uranium will continue to be the valuable target metal it was when ion exchange for uranium began to be investigated in the 1970s. The slackening of the growth of nuclear power programmes may reduce the rate of increase in demand such that alternatives to land-based resources never really become attractive propositions. If uranium ceases to provide the impetus for ion exchange resin development it is unlikely that, by themselves, the other industrially valuable metals would justify the same research commitment; although, once a technology has been established for uranium, it might be adapted for other metals.

If the world-wide future for uranium demand is unclear at present, at least Japan has no doubts that it should pursue the extraction of uranium from seawater regardless. Japan's need to secure its energy supplies is well-known and so Japanese researchers have been interested in a range of potential new sources, including ion exchange. The development of nuclear power has been central to Japanese energy policy since it was seen, in the long term, to offer a better guarantee to supply than fossil fuel power generation options. Even so, domestic supplies of uranium meet only one per cent of demand and no large increase in the supply is foreseen. Therefore, initially at least, the supply of uranium is a matter of great strategic importance to Japan. Medium-term supplies are covered by contracts with Canada and Australia, but Japan is intent on reducing even this degree of dependence on other countries (particularly, perhaps, since uranium supplies might be used by the Western industrial nations as a weapon to threaten Japan if it continues to make incursions into Western markets). So Japan, more than any other country, may be prepared to pay a premium in order to secure uranium.

Japanese research began in the 1960s, but the programme accelerated in the mid-1970s. Ion exchange was not the only method considered, but it has become an increasingly important aspect in the last few years. The Japanese programme has set out to build a pilot plant. Originally, it should have been in operation by the mid-1980s, but it will probably not be in commission until near the end of the decade. This plant is only intended to produce 10 kg. of uranium per annum, but its cost is estimated at about $US 20 million. If the pilot is successful then a plant producing a few tonnes per year is planned before a full-scale plant with an output of 1,000 tonnes per annum comes into production in the early years of the next century. This commercial plant would either be sited with an OTEC scheme or would utilize the Kuroshio current in order to achieve high flow rates and ensure the replenishment of the seawater.

There have been delays in this Japanese programme and its future is not clear, but the commitment to uranium self-sufficiency is great enough to suggest that the programme will continue for a while yet. However, the Japanese project does not reflect the commitment elsewhere where the strategic requirements of the 1980s do not require such a premium to be paid in order to ensure supply. Ion exchange does offer a better prospect for the retrieval of low-concentration but high-value metals from seawater than that hitherto possible with conventional technologies. If the scale can be made practical and a resin produced that will withstand the requirements of continual commercial processing then uranium and eventually other metals may be extracted from the seas.

COMMENTARY

With the obvious exception of sand and gravel dredging, the marine technologies described above are of a large scale and high technological sophistication. They appear to have more in common with manganese nodule mining than with dredging for phosphorites. The current states of resource availability and estimates of world demand do not hold out much promise for the immediate commercial development of hydrothermal sulphides, radiolarian oozes or the extraction of metals from seawater. Indeed, it appears that, certainly in the case of oozes and metals from seawater, these technologies only have a future in conjunction with other marine engineering operations. If the future brings integrated ocean technologies then the prospects for the extraction of metals from seawater are more encouraging since the latter may be tenable in association with an OTEC plant or desalination scheme or a tidal power development, where the capital costs of construction and the operational costs of pumping enormous volumes of water can be spread between two programmes. However, in these circumstances the future of both oozes and sulphides is much less rosy since the only deep sea mining technology with which they might link is manganese nodule mining, the immediate prospects for which do not imply rapid development in the next decade or so.[10]

In the short term the extension of the operations to extract minerals from the Dead Sea is far more likely and the current viable potash production may provide a framework within which to develop processes able to extract more valuable industrial metals from the high concentrations in the Dead Sea. Processes developed initially for the unique conditions of the Dead Sea might be subsequently refined to operate at lower concentrations. It is unlikely that such processes (based almost certainly upon conventional extractive procedures rather than upon ion exchange or any other more exotic alternative) would be applicable for a wide range of metals but, given sufficient development and other favourable circumstances, in specific situations they may be useful.

The current climate of resource economics does little to enhance the prospects for any of the more exotic programmes. Extrapolations of rapidly increasing world demand for a range of raw materials made in the late 1960s, and upon which many of the marine programmes were predicated, have not been fulfilled and in many instances new land-based resources have been able to meet the lower growth rates of demand for those raw materials. Even if accurate costings could be produced for marine mineral extraction (and such costings are not possible given the uncertainties of the programmes), it is not likely that the prices at which materials could be delivered onto the world markets would be competitive enough with traditional supplies to guarantee a commercial future. The extent to which individual countries, other than Japan, consider strategic needs crucial enough to pursue some of these projects in spite of the high cost of the final product has yet to be seen.

The schemes discussed in this chapter are the Cinderellas of marine mineral exploration. For the reasons already outlined, they do not appear to have any future in their own right, only in conjunction with other schemes. The technological requirements for them are no more difficult to meet than those of several other proposals which are far more likely to be realized as commercial enterprises: if an OTEC plant can be built and operated then almost any other marine technology ought to be within the bounds of the feasible. But this illustrates most graphically that technical capability is insufficient to guarantee development and in these cases the product on offer is neither politically nor economically of so high a priority as to provide the impetus to overcome the technological and engineering problems of a novel project and to ensure development. The fact that the future of these programmes may only be viable as an adjunct to other schemes does not mean that they can be ignored or classified as being unimportant. They are serious proposals, which in the climate of the late 1960s and early 1970s seemed to be part of a resource strategy that any prudent industrial nation must pursue in order to ensure supplies of important raw materials. In pursuit of such security of supply, no proposal should be rejected until it is shown to be wholly untenable. It may yet be that the future of many new marine technologies is best served by an integrated approach in which complementary operations are sited together in order to gain the benefits of economies of scale for construction and running costs. In these circumstances oozes, sulphides and metal extraction from seawater may all become very important.

NOTES

1. See Chapter 7 below.
2. Recolonization of the disturbed area may be speeded up slightly since, in stirring up the sediment, nutrients are made available that had previously been buried beneath the sea-bed.
3. It will probably already be clear that the origin of polymetallic sulphides is similar to that described in Chapter 3 for processes that produce the Red Sea metalliferous muds. Both in terms of origin and mineral composition, there are similarities, but the Red Sea deposits are of relatively unconsolidated muds and so are amenable to dredging and pumping. The Pacific Ocean sulphides could not be lifted by the techniques advocated for the Red Sea deposits.
4. As was observed in Chapter 2, other circumstances have changed since the

early enthusiasm for manganese nodules and the absence of the international legal framework is not necessarily the only reason for the lack of progress in the last few years.

5. Other deposits may also lie close enough to Hawaii to be claimed as being under American jurisdiction.

6. In Western Europe only France has shown any serious interest and she may soon begin surveys in the Pacific.

7. Bromine and magnesium are two of the few elements that have been extensively extracted from seawater.

8. Of the more exotic proposals, there is one that uses the property of some marine micro-organisms to concentrate certain metals. Possibly with bio-technological development, organisms could be produced which concentrate specific target metals. This is not a proposal of any immediate relevance.

9. The concentration of uranium in seawater is of the order of three parts per billion.

10. Whether development is carried out on their own or in conjunction with manganese nodule mining, it is important to remember that both hydro-thermal sulphides and radiolarian oozes will require the establishment of a political and legal framework before a full commitment to development can be expected. It has been the lack of such a framework that has done much to stall the manganese nodule programme.

5 Ocean Thermal Energy Conversion

INTRODUCTION: THE CONCEPT

The search for a cheap and abundant source of energy is not a new one, but since the Second World War the emphasis in this search has been on renewable sources of energy. More recently still, the desirability of environmentally benign energy production technologies has been realized. Not surprisingly, many proposals for renewable energy projects have looked to the oceans, particularly towards their movement; tides, waves and currents. None, apart from some notable tidal power projects, have impinged significantly on the energy demands of the world; however, one of the most interesting proposals does not see the ocean's value in its motion but in its temperature or, to be more precise, in the temperature differences within the body of the ocean. It is called Ocean Thermal Energy Conversion (OTEC) and throughout the middle and late 1970s it formed the vanguard of research into renewable energy systems from the sea. Although there has been a relative decline in interest in the United States, elsewhere in the world research and development continue and many observers believe it will yet prove to be an important component in energy supply. For this reason alone the case of OTEC is instructive, but it also illustrates general issues which are faced by all projects of ocean resource development. In particular we shall see the sheer scale of commitment required if many of these projects are to have a significant impact on the availability of energy or minerals, or even food. No other ocean technology programme, starting from scratch, has been envisaged on the scale suggested by OTEC proponents.

OTEC is a solar energy system: the sun warms the ocean's surface, which becomes a reservoir of heat. The principle of OTEC is to extract that heat and the concept is thermodynamically very simple. It is, perhaps, deceptively simple, since, although easily understood, it belies the immense complexity and expense of turning the concept into a working, viable power station.

Energy can be obtained wherever a temperature difference occurs. In a conventional power station, say one that burns fossil fuels, combustion provides the high temperature to generate steam and the relative coldness of the environment can be used to condense the steam back to water. The difference between the two temperatures involves the water in a change of state and in the steam phase it can turn a turbine to produce electricity. In

OTEC the principle is the same, except that the temperature differences in the cycle are those between warm seawater at the ocean's surface and the very cold water at 1,000 m. deep. If it were possible to design a system to utilize it, this temperature difference could be used to power a turbine. In the tropics the surface temperature of the sea is between 20 °C and 25 °C, while the Arctic and Antarctic bottom waters, which circulate at a depth of 1,000 m., are at a constant 4 °C.[1] Hence, in favourable locations, differences of over 20 °C are available (see Figure 5.1).

Clearly, the temperature difference for OTEC is very much smaller than in a conventional power station, where it may be, typically several thousand degrees. This is reflected in the lower thermal efficiency of energy conversion in OTEC,[2] but the potential total resource is so enormous that the relatively low efficiency is not important.

So the principle is simple and it is not difficult to outline the design for power production cycles to extract energy from the temperature differences available to OTEC. There are two broad categories for the system. The first is directly analogous to the steam-generating systems of con-

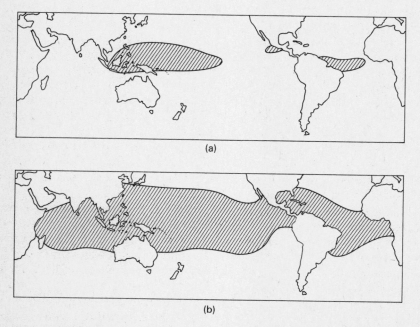

(a)

(b)

Figure 5.1 World-wide distribution of OTEC Thermal Resource, showing resource (a) between surface and 500 m. depth, and (b) between surface and 1,000 m. depth. Shaded areas have monthly average change in temperature greater than 20°. *Source*: U.S. Department of Energy.

ventional power stations in which a liquid is boiled to produce a gas which will expand through a turbine set to produce power. The gas is then condensed back to liquid form in order to repeat the cycle. This is a 'closed cycle' and such a cycle is feasible for OTEC, except that the fluid would not be water but rather a liquid that would boil between 4°C and 25°C at the working pressure in the system. There are several fluids which could be used, including ammonia, propane and some refrigerants. The warm water from the ocean's surface heats the fluid past its boiling point, the expanding gas drives the turbine and the cold water from the ocean depths re-condenses it.

The alternative system does not use a separate working fluid. In the 'open cycle' the warm seawater is evaporated at reduced air pressure to produce a low pressure steam which directly drives a turbine before recondensing at the cold water sink. There are some advantages to the open system, but in general most proposals for OTEC power production have chosen the closed cycle, and we can use that system to illustrate the major components of an OTEC plant. It would need pumps and piping for the inlet and exhaust of both warm and cold water as well as plumbing for the working fluid circuit. Also, it would require an evaporator and a condenser where the working fluid and seawater circuits intersect, and a turbine with associated electricity-generating equipment. The details of how these components might be fitted together vary between individual designs, but the general outline is shown in Figure 5.2.

Broadly speaking there are three distinct approaches to the question of how an OTEC installation would be realized and operated. One proposal is the 'plant ship', which is conceived as a floating chemical works producing materials that involve energy-intensive processes. Ammonia, aluminium and hydrogen have all been suggested as products which would benefit from cheap electricity available on-site. The commercial viability of any plant ship proposal would depend upon the specific economies of each case and more will be said of this later, but such proposals do have one very attractive feature. Using the electricity at sea obviates the need to transmit it back to shore along marine cables tens, if not hundreds, of kilometres long. This solution bypasses one of the most serious technical problems of OTEC.

A second option is for a floating platform station to feed electricity directly into the grid of a nearby on-shore area. This is the OTEC concept in its classic form, but all the most serious technical problems must be solved before the system is feasible, and these problems are considerable.

The third form is something of a compromise. It is a land-based installation which obtains its warm and cold water from long pipes extending beyond coastal waters into the deep ocean. Such a system is more easily achieved with existing technologies than either of the platform proposals.

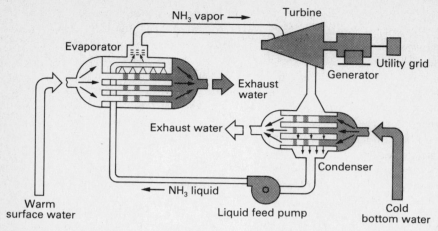

Figure 5.2 Schematic layout of OTEC system. (Adapted from TRW's 'OTEC, Solar Power from the Sea' (1980)).

Although the land-based station does not represent the full potential of OTEC development, it may well prove to be the most hopeful possibility for commercial exploitation in the medium term.

AMERICAN DEVELOPMENT

Until the end of the 1970s, the United States was carrying out the largest OTEC research and development programme. The expenditure on it rose from $85,000 in 1972, to $8.5 million in 1976 and $38.5 million by the end of the decade. This increasing financial support for OTEC obviously reflects the United States' response to the problems of oil supply and price in the early 1970s, but it should be remembered that these factors spurred research into many alternative energy sources. There were other features of OTEC which made it especially appealing and which retained support for it as other renewable energy programmes fell from favour. Strategically, American energy policy in the 1970s had two aims. In the medium term it sought to reduce dependence upon imported oil and in the longer term it needed to secure sufficient energy supply once oil reserves became depleted. In other words, any plausible long-term substitute for oil had to be feasible on a very large scale—on a scale sufficient to meet a significant proportion of American demand. OTEC could do this because each OTEC station was expected to produce electricity at rates comparable with individual fossil fuel (or nuclear) power stations. For example, it was proposed that the first generation of commercial OTEC stations would be rated at about 500 MWe (1MW $=10^6$ watts) each and one survey estimated that an array of such stations in the Gulf of Mexico could generate up to 30 GWe (1GW $=10^9$ watts). At present levels of consumption, this represents about one-third of the demand on the United States South/South Eastern power grid. It was this potential for large-scale energy production (together with its relative independence of transient weather changes)[3] which made OTEC a candidate for producing base-load electricity. Furthermore, if OTEC could replace conventional power stations, it could also be an alternative to nuclear power. This did no harm in broadening OTEC's political support since one of the major arguments made in favour of the nuclear option was that no other proposals were practical on the scale of nuclear energy; that was until OTEC, which offered a non-nuclear strategy without implying a reduction in energy supply.

In summary, during the later years of the decade, OTEC promised enough to gain at least passive support from a range of groups, institutions and government agencies, and this permitted the active proponents of the project to proceed with a strong campaign. The success of that campaign is illustrated by the passage of two bills through Congress, which received wide political support, for the reason already stated, but which were also helped by the active support of Congressmen from states that would benefit from OTEC power. In particular, Hawaii, which has few indigenous fuel resources but does have access to an ocean with the appropriate thermal profile, was exactly the type of site where OTEC plants might be first deployed commercially and so, with the energetic backing of the Hawaiian Congressmen, the two bills were passed.

The first bill was the 'Ocean Thermal Energy Conversion (Research and Development) Act'[4] which recognized the need for alternative energy sources and which stated, in particular, that 'ocean thermal energy is a renewable energy source that can make a significant contribution to the energy needs of the United States' (HR 7474, section 2 & 3). The Act called for a clear programme of OTEC development which was to demonstrate a 100 MWe system by 1986, a 500 MWe system by 1989 and, by the mid-1990s, be able to generate electricity at an average cost competitive with conventional fuels. A total output of 10,000 MWe was the target by the end of the century. In pursuit of this programme, provision was made for a budget of $75 million for the financial year 1982 on OTEC research.

The other legislation passed was the 'Ocean Thermal Energy Conversion Act'[5] which laid the legal and administrative infrastructure required for OTEC to proceed towards commercialization. It included the provision for licencing OTEC stations; it delineated the framework of regulations relating to the operation and control of OTEC installations. But most crucially it provided for $2 billion of assistance, in the form of loan guarantees, through the 'OTEC Demonstration Fund' in order to encourage the continual advance towards pilot plant demonstration of OTEC's commercial viability.

Both pieces of legislation passed through Congress with very little opposition and were signed by President Carter during 1980. They represented the culmination of an OTEC programme, that had already been building up for nearly a decade, by establishing a legal and financial environment for the project which demonstrated governmental commitment to its future. However, that period now seems to have been the apotheosis of OTEC, for although still important, it has relatively declined under the Reagan administration.

Since 1980 the emphasis of American energy policy has changed. In keeping with the overall philosophy of the administration, the commercial development of OTEC is to be the responsibility of industry, not government. Federal financial support for OTEC was reduced in the 1981 budget proposals and the participating corporations have been expected to put up their own money for the next stages of development. Although both government and individual firms have committed extensive resources to OTEC so far, both appear to believe it is up to the other to show faith in the commercial viability of OTEC and to produce the necessary finance. The result of this has been stalemate. Another factor is the expectation of the participating firms as to the most likely sources of federal R & D contracts. Firms such as Lockhead and TRW are primarily concerned with defence and aerospace high technology and they may have anticipated an increase in American defence expenditure under Reagan and a consequent increase in defence contracts. Hence, the firms may have felt there was less reason to remain committed to OTEC if more money was going to be available in other, more conventional research.

Overall, the effect has been a dramatic slowing down of the pace of the programme and only two projects seem likely to be realized in the near future. The US Department of Energy is still financing design proposals for a 40 MWe land-based plant and a 40 MWe shelf-mounted plant for Hawaii. But probably the best chance is for a plan to build a 48 MWe land station on the island of Gaum by a consortium of American companies. But both the Gaum and Hawaiian projects are a far cry from the proposals to create an extended OTEC array in the Gulf of Mexico. We will see later that the United States is not alone in lowering her expectations as to the scale of feasible OTEC first-generation plants, although the American change in policy is more by default than by any rational reassessment of the aims of the programme.

The principle of OTEC was simple: schematic designs had been proposed and financial support was available from government agencies. All that was required was the small matter of actually building the plants. This, of course, is no mean task and involves two types of technical challenge. On the one hand it requires the scaling-up of otherwise well-known technologies in pumping, heat exchanger design and underwater electrical power transmission. While on the other hand it also requires some radical developments for the construction of the platform on which to float the OTEC plant, for the design and construction of the pipe to bring up the cold water and in the knowledge of materials' performance in marine environments. Problems of both types will have to be solved if OTEC is to have a commercial future. To that end several development projects have been proceeding under the 'OTEC Engineering Research and Technology Development Program' funded mainly by the U.S. Department of Energy.

Several companies and institutions have been involved in the design and construction of equipment. The two major participants so far have been Lockheed Missile and Space Company Ltd. and TRW Defence and Space Systems Inc., both of which have extensive experience of government financial R & D particularly in defence and aerospace contracts. Both firms, along with the Applied Physics Laboratory at John Hopkins University have produced designs for components and complete OTEC systems.[6] No simple design is obviously cheaper or more efficient and no single material meets the fabrication and maintenance requirements in all aspects. It is still a matter of testing and engineering trade-offs before final designs can be selected which offer the best overall performance.

To consider the technical progress that has been made on OTEC (and the difficulties which still remain), it is useful to break the system down into its major sub-units and deal with each in turn.

The function of any form of free-floating OTEC station (whether a plant ship or one generating electricity for transmission to an onshore power grid) will be twofold. Not only must it house the OTEC hardware and provide crew accommodation, but it is also the mounting from which the 1,000 m. pipe to the cold deep ocean water must be suspended. The platform will be subject to wave and current action, hence designs seek to minimize surface stresses on the platform and the cold water pipe. A variety of designs are proposed from variants on a conventional ship's hull through to totally submerged spheres (see Figure 5.3 for some examples). Lockheed have advocated a spar buoy shape and John Hopkins Applied Physics Laboratory proposes a rectangular semi-submersible; both of these ought to achieve substantial control of the stresses on the unit. TRW and Sea Solar Power, however, both have chosen surface vessels which may be more immediately feasible than the semi-submersible designs. In fact, one of the test rigs used for early demonstrations, OTEC-1, was based on a floating hull shape and although it possessed good stability there is doubt over how such surface vessels would perform when attached to a full-size cold water pipe.

The uncertainty over the platform is predominantly one of design. Its

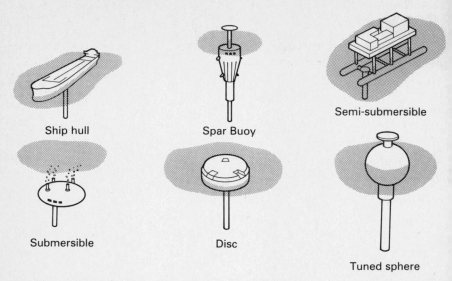

Figure 5.3 Possible platform configurations. (Adapted from Marine Technology
Society's *Marine Technology* 80).

construction would not be very different to existing practice in the
construction of offshore installation for the oil industry. For example,
displacement tonnages for a 400 MWe OTEC station have been suggested
in the range of 200,000 to 800,000 tonnes and the Dunlin A platform in the
North Sea has 800,000 tonnes of displacement. So the size of the OTEC
platform is not, in itself, a major problem. Any of the configurations could
be constructed in steel or concrete. Present designs tend to favour steel,
although this reflects American oil industry experience where steel rather
than concrete is still the primary material for offshore oil rigs. The likely
combination for a first-generation OTEC plant would therefore be a
surface floating (though probably not hull-shaped) steel platform strongly
influenced by construction practices in the oil industry. More exotic
designs in concrete may be later developments.

If the platform does not involve any major obstacles, then the cold water
pipe (CWP) provides plenty! In a 400 MWe OTEC station a pipe 30 m. in
diameter will hang vertically to a depth of some 1,000 m. below the
platform. The rate of water flow up this pipe has been compared to that of
the River Nile! There are numerous problems associated with the CWP. It
will be subject to several different forces; forces caused by its own weight
and forces caused by water movements at different depths as well as the
forces at the connection with the platform. These forces have not yet been

fully quantified, neither has the design been settled. The CWP is one area of OTEC development where there exists no comparable experience which can be scaled-up to meet OTEC needs. Hence, a diversity of solutions has been offered. Both rigid and compliant materials have been suggested for construction: reinforced concrete, steel, glass reinforced polymers and nylon reinforced rubber. Single pipes and clustered designs both appear feasible.

However, whatever design and materials are selected, the physical deployment of the CWP beneath the platform remains a problem. If constructed on land, it would be vulnerable to damage during transportation to the OTEC site, alternatively, assembly *in situ* (from prefabricated sections perhaps) could be very difficult as well. The sheer size of the CWP presents the main problem and there is no experience on which to draw to solve the problem for construction, transport and deployment.

In 1980 a report by the United States Office of Technology Assessment noted that although important advances had been made in many aspects of OTEC development, the programme to produce a suitable CWP was well behind and required a substantial effort if the situation was to be improved.

In the two OTEC test rigs the CWP was made of polyethylene. On Mini-OTEC a single pipe, 24 in. in diameter and 2,000 ft long performed satisfactorily for two months during the summer, while on the later OTEC-1 there were three 48 in. diameter pipes clustered together around a plumbline for stability. This array was around 2,000 ft long. Although polyethylene and these designs are thought to be suitable for generating plants of up to 10 MWe, it is obvious that there is a considerable gap between this and the requirements of a system appropriate for the 400 MWe plant.

The CWP is seen as the single largest technical problem obstructing the rapid development of OTEC. So, not surprisingly, many observers are beginning to see the immediate future for OTEC in onshore installations, for which the CWP would lie on the sea-bed. In this case the total length of pipe might be several times greater than one dropping vertically from a platform, but that might be a small price to pay for sidestepping the problems of the CWP and floating platform configuration.

Both floating and onshore installations have the same requirements regarding the internal plumbing for OTEC. The components most crucial to the efficiency of the sytem are the heat-exchangers, of which there are two; one in the evaporator and one in the condenser. Their performance is critical since they must transfer as much heat as possible from the warm water to the working fluid (the ammonia, for example) in the evaporator and as much as possible out of the working fluid back into the cold water at the condenser in order to maintain the greatest possible temperature difference within the cycle. In other words, the heat exchanges need to be so efficient at this heat transfer between water and working fluid as to make the temperature difference within the cycle as near as possible to that actual temperature difference between the warm and cold seawater. The techniques used to maximize the efficiency of heat exchanges are well-known. Usually, the design aims to achieve the largest surface area of contact between the heat-exchanger and the two fluids. Several well-tried designs can provide high surface area to volume ratios (see Figure 5.4), but for OTEC purposes even higher heat transfer efficiencies are sought. Some of it can be gained through control of the fluid flow in the exchangers by a smooth, slow flow along the exchange surface, but by itself this is still not sufficient. What is needed is what is called 'enhancement', which is a method of increasing the effective surface area of the exchanger components. This could be achieved by roughening the surface so that at a microscopic level the ridges and furrows of the roughened surface provide an increase in the available area, but the preferred method is to coat the surface with a porous, thermally conducting material, the internal structure of which provides the increase in surface area.

By means of these three measures (maximizing design, fluid flow control and enhancement), heat exchanges of twice the usual efficiency have been devised. Naturally they are expensive, but no OTEC problem is likely to be solved cheaply. However, heat exchange efficiency is not the only problem here. One side of the exchanger will be in contact with seawater and biofouling of the surfaces by micro-organisms must be prevented since it would ruin the heat exchange properties of the components. Some form of cleansing is required to remove biofouling. This could be either a chemical or a physical process, but if the problem is not solved no amount of enhancement will be successful.

As with other OTEC components, the heat-exchanges will be large. For the 400 MWe system effective surface areas of one square mile have been suggested as the requirement. This is beyond current common industrial practice.

The choice of material for the exchangers is not straightforward either. The traditional materials for marine heat exchangers have been stainless steels and cupro-nickel alloys, but if ammonia is the chosen working fluid cupro-nickel will be unsuitable since it is incompatible with ammonia. Aluminium would be compatible with ammonia, but its resistance to marine corrosion is not good. Titanium would be compatible with the ammonia; it has excellent resistance to marine corrosion and its strength-to-weight ratio permits the design of very thin walls to aid heat exchange. However, titanium is expensive, it is not as easy to process as aluminium and it is not available in the quantities needed for OTEC.[7] If another working fluid were chosen then a new selection of options would be generated, but none are any more simple than the ammonia example.

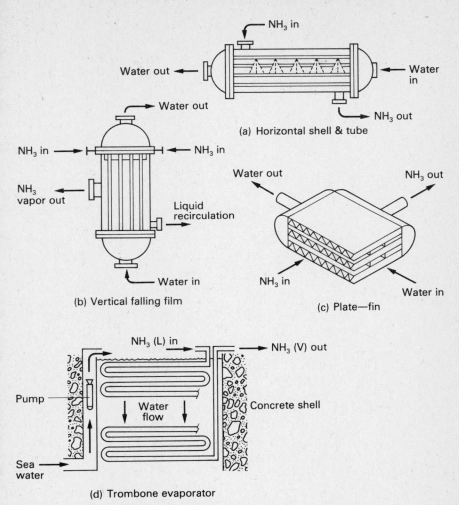

Figure 5.4 Heat-exchanger types. (Adapted from *Quest*, **3**, 1979).

In fact, in OTEC tests so far, the ammonia/titanium combination has been used. On OTEC-1, heat-exchangers 1 MWe in size have been constructed by TRW in titanium with the enhancement on the ammonia side of a shell and tube arrangement. The shells are 50 ft long, 10 ft in diameter and contain 6,000 tubes. It is instructive to realize that this is for 1 MWe; i.e. 1/400 of that needed for the proposed commercial plant! Other systems have been tried in OTEC-1, including a plate and fin design in stainless steel with enhancement of a porous aluminium surface coating. The latter design was costed as slightly cheaper than the titanium shell and tube exchanger, but uncertainty arises over its longevity in the marine environment.

The development of the heat-exchanger is typical, not only of the technical problems faced by OTEC, but it is also characteristic of many novel marine projects. Experience of the marine environment remains limited.

The platform, CWP and heat-exchangers do not exhaust the technical problems of OTEC, but they are the major ones. No one has yet built a turbine with the characteristics required for power generation from OTEC, but there is little disagreement that it is within the potential of current practice. Similarly, the transmission of electricity from a floating offshore OTEC plant is now considered to be within reach. For instance, laying a 250 MW direct current cable across the Norwegian Trench in 550 m. of water is comparable to the requirements of OTEC. The importance of electricity transmission from OTEC is primarily economic rather than technical since the cost of transmission will be a significant factor in determining the price of OTEC-generated electricity, and if the cost of transmission is too high it may act as a further incentive for the development of land-based installations first, as these do not require underwater electrical transmission.

The final area in which OTEC needs exceed common practice is in pumping,[8] where the cold water has to be pumped up from the depths to the condenser. The total pumping capacity required is far beyond that of existing seawater pumps, but combinations of pumps should be suitable. The main problem will be to maintain constant pumping rates, particularly on a floating platform that is rising and falling in heavy seas, which would cause the pressure-head of water to fluctuate.

This summary of the American OTEC programme demonstrates the technical problems of OTEC development and the advances that have been made towards solving them in the last ten years. It is clear that some major ones still remain. Although the American experience well illustrates the state of OTEC research, it is incorrect to assume an American monopoly on OTEC. Other countries have become interested in it and their approaches are not always the same as those of the United States.

OTHER PROGRAMMES

Since the relative decline in American commitment to OTEC research, programmes in other countries have assumed increased importance and now Japan, India, Holland, France and Sweden have all begun projects.

Japanese interest is a long-established one. The country's dependence on imported fuel[9] has encouraged an open-minded attitude towards new power sources. In the early 1970s investigations into a variety of novel energy-generation methods began; one of the proposals studied was for OTEC. Then, in April 1974, the Ministry of International Trade and Industry began to fund OTEC research in collaboration with industrial and academic institutions. A programme was prepared that sought to demonstrate, step-by-step, the feasibility of OTEC. The initial goal was to design a small, experimental land-based station and to build towards the goal of a 100 MWe floating power plant. In this programme the Japanese have faced many of the same technical problems as in the American development: the choice of materials, the design of heat-exchangers, the platform and CWP design and the assessment of the available thermal resource. The Japanese choices have not always been the same as the American, but the two programmes have produced not dissimilar designs. There is one obvious difference, however. The Japanese project has aimed to design and operate firstly a land-based station and then a 100 MWe floating one. This contrasts with the initial American aims of commercial stations of 400 MWe at sea and suggests that the Japanese will have their OTEC plants operating commercially before the United States. Indeed this does appear to be the case. In 1981 the Japanese operated a 120 KWe pilot plant on the Republic of Nauru. This land-based plant has produced a net output of 31.5 KWe, which although not very great, does represent considerable success for the programme. On the basis of this demonstration the Tokyo Electric Power Services are building a 2.5 MWe land-based plant for the Nauru government. This should come on-stream in 1986 and would be the first commercial OTEC plant. These successes in small-scale OTEC have created far more interest in the tropical Third World than the much grander aims of the United States programme.

After the United States and Japan, the next major participant is, perhaps surprisingly, France[10] which, although it does not possess tropical oceans adjacent to its own coastline, does have strong ties with its ex-colonies in the tropics. Through the French government's energy department, CNEXO, small projects are proceeding in French Polynesia and on the Ivory Coast. These schemes are intended to create plants of not more than 40 MWe since the French designers believe this to be the maximum size for which a CWP is available within current technology.

Similarly to France, the Dutch and Swedish programmes are small in scale and are undertaken in association with tropical countries. The Indian programme is very new, but with its long coastline of tropical oceans its programme of OTEC development for domestic use may become one of the most important.

One of the recent entrants to the OTEC field is Great Britain. A consortium of British firms may soon be constructing a small floating plant in the Caribbean. It is no more unlikely for Britain to pursue OTEC than it is for any of the other European nations with tropical ex-colonies. Indeed, the parallel between OTEC and some aspects of offshore oil exploration has already been noted. So, in many matters of common practice between oil exploration and OTEC, the British experience in the North Sea may be invaluable to a British programme. The experience may be important for the construction of platform and CWP components, for the logistical support of extensive offshore operations, or simply by creating the confidence to tackle a new marine technology at the limit of current knowledge. The development of North Sea oil production has been a very considerable achievement and much of the skill and experience has accrued to British firms. As the rate of development in the North Sea slows down, these skills may find a new outlet in a British OTEC project.

THE FUTURE FOR OTEC

The current interest in small-scale land-based OTEC installations suggests that, by the end of the 1980s, there will be in operation OTEC power stations generating perhaps up to 10 MWe for industrial and domestic consumption. As to whether OTEC will ever be deployed on the scale initially proposed in the United States, it is difficult to assess. Even if the remaining technical problems of the design and construction of 400 MWe floating stations were solved, this in itself would not be sufficient to ensure the commercial success of OTEC. Yet, if OTEC is to be significant to the power supply of a country such as the United States, the energy it generates must have an average cost that is comparable to that of conventional power stations. Economic competitiveness is not a necessary condition for small-scale OTEC development. As we have seen, many of the countries now interested in OTEC have no conventional fuel reserves of their own. In order to develop OTEC as an indigenous energy resource, these states are not overly concerned if the short-term cost of OTEC-generated electricity exceeds that of importing oil or coal. The prospect of being independent of imported fuel largely outweighs the narrower economic arguments against OTEC. However, in the United States, which is relatively rich in fossil fuel reserves, which has powerful political lobbies for both fossil fuel and nuclear power, and which has such a high demand for energy that cost can never be ignored, the economic competitiveness of OTEC must be demonstrated if an extensive commercial programme is to continue.

Unfortunately, there is little agreement as to the likely commerical economics of OTEC on a large scale. The American Office of Technology Assessment has reported on the progress of OTEC and reviewed the cost estimates produced by the various consortia. The cost of producing electricity from OTEC is almost entirely due to the capital costs of construction and operation: The raw material fuel (i.e. seawater) is in effect free. Therefore, until accurate estimates of capital expenditure are available, there must remain uncertainty over the average busbar cost of OTEC electricity.[11] A whole series of assumptions have to be made as to the cost and performance of each component, the average temperature differences available to the station, interest rates on the capital cost borrowing and the capacity factor (i.e. the proportion of time that plant is operating and not undergoing repair or maintenance). Not surprisingly, a wide range of estimates has been produced! Based on late 1970s prices, estimates of the cost of OTEC electricity varied from one half to four times the cost of fossil or nuclear-generated electricity. The major design groups tended to be optimistic and estimated in the range of 0.5 to 1.5 of the cost of conventional generation,[12] while independent assessments lay in the range between twice and four times those of conventional power stations. Until there is more firm evidence to narrow the range of the individual parameters, it is impossible for the economics to become any clearer.

Another imponderable is the international price of oil. Largely as a result of the rapid price increases by OPEC in the early 1970s, the economics of OTEC began to look more attractive as the price rises for oil were extrapolated into the 1990s. Subsequent experience has not borne out this prediction. A world glut of oil has resulted in a fairly stable price recently, but either an increase in demand caused by industrial recovery in the West or a threat to supplies from political crises in the Middle East could easily send oil prices rocketing and OTEC might become financially more attractive once again. It is difficult to anticipate oil prices even a few years ahead; to attempt to estimate them for fifteen to twenty years ahead therefore is impossible. The comparative economics of OTEC could change considerably, even for large-scale programmes such as the Gulf of Mexico array.[13]

Financial considerations, like technical ones, are a necessary, but not sufficient, element in ensuring the wide diffusion of a new technology. Any new large-scale technology must now face scrutiny on its environmental impact. OTEC is no exception to this. Much of its early support in America was due to its presentation as a renewable energy source, an alternative to the depletion of fossil fuel reserves, and an alternative to an increased programme of nuclear power. Furthermore, it appeared to have no ecologically damaging side effects. Certainly, on a small scale it has virtually no environmental impact. But on a large scale some questions are raised: primarily the matter of thermal pollution. Large quantities of cold water will be discharged at the surface of the ocean. This water will have warmed up beyond its initial 4 °C, but it will be much cooler than the surrounding water. The rate at which the cold plume will disperse is unknown. If it is slow, then significant cooling of the upper layers of the ocean may occur.

This could have several repercussions. Localized changes in sea temperature might affect the micro-climate downstream of the OTEC plant and disturb the thermal balance of the surrounding ocean. Also, it might be self-defeating since any reduction in the average surface temperature will reduce the efficiency of the OTEC cycle. All these problems would be exacerbated in a basin such as the Gulf of Mexico which has restricted water circulation. Certainly, any proposal for a large deployment of OTEC in the Gulf of Mexico has to deal satisfactorily with this question, as indeed it must with another problem if many plants are to be sited in a confined area. The importance of preventing biofouling in the heat exchangers has already been mentioned. One method of achieving this would be to flush out the system with an agent such as chlorine. There could therefore be a problem from the discharge of chlorinated water, especially from a number of plants in a small area. Even in low concentrations this would be undesirable and its impact on marine life is unknown. It illustrates once more that even an apparently benign technology such as OTEC can have unanticipated consequences for the oceans if deployed on a large scale.

COMMENTARY

The history of OTEC development is a most interesting and instructive case of new marine technology. It is a renewable energy resource which is not incompatible with the needs of a highly industrialized and energy-hungry society. Thus, governments and contractors have conceived OTEC on a very large scale; this alone makes it unique among the new marine technologies. No other marine programme has been backed by the financial and political support given to OTEC by the United States in the late 1970s. Under that weight of effort, considerable progress was made on many fronts, with possibly the cost of the heat-exchangers and the construction and deployment of the CWP remaining as the major obstacles to technical success. But as we have seen, technical virtuosity alone cannot guarantee the outcome. The withdrawal of American political commitment to OTEC has severely damaged the programme, which now, together with projects in other parts of the world, is aimed at a much less ambitious scheme. This may not, paradoxically, damage the long-term prospects of OTEC. The penalties for being too ambitious in advanced technological projects can often be worse than those of undue caution. This is especially true of marine technologies, which pose challenges nearly as great as those faced in space exploration.[14] The present state of the OTEC art suggests that in the 1990s there will be land-based OTEC stations supplying a few tens of megawatts each. These will be in specific suitable sites where OTEC provides the appropriate technology to utilize the only indigenous energy resource. It will not necessarily, either in its scale or its economics, be applicable to more widespread exploitation in that form. But there will be the opportunity for piecemeal developments to the system which may, in time, lead to the realization of 400 MWe floating plants.

NOTES

1. The constant temperature of Arctic and Antarctic waters leads to the possibility of a 'reverse OTEC', where the water at the poles is the heat source and the air, which can be very much below freezing point, is the cold sink. The air temperature in the polar regions can be so low as to give much greater temperature differences than those available to the tropical OTEC system.
2. For example, Carnot efficiency of a conventional power station would be of the order of 80 per cent. Whereas OTEC would be (300 − 280)/300 approximately equal to 7 per cent. These Carnot efficiencies are the theoretical maxima.
3. The heat reservoir in the oceans is so great that temperatures are relatively stable from season to season in the tropics, hence diurnal changes are insignificant and even several days without sunshine would not affect output. This is in contrast to most other renewable energy sources which, for example, require direct sunlight or a minimum wind velocity to be operational.
4. See HR7474 and S1830.
5. See HR 6154 and S2492.
6. There are three major participants, though there have been many others involved at various stages. For example, the Westinghouse Electric Corporation has produced several power system designs for OTEC, including open cycle designs for which they claim great advantages over the closed cycle. Also, a firm named Sea Solar Power has been active in OTEC for many years and, although not directly involved in the construction contracts, Sea Solar Power do own many patents which may prove crucial to OTEC's future.
7. One estimate shows that to build eight titanium heat-exchangers, 400 MWe in size, it would take around one-third of total American titanium output.
8. There are other pumping requirements in OTEC, such as circulation of the working fluid and the intake of warm surface seawater, but neither of these pose any problem.
9. Japan has to import between 85 and 90 per cent of its fuel consumption.
10. The French interest in OTEC does have historical precedents. The idea was first proposed by a Frenchman called d'Arsonval in 1881, and in the 1930s a French engineer, Georges Claude, had some limited success in demonstrating the concept with a pilot plant in Cuba. Claude's main problem was, interestingly, the CWP!
11. The busbar cost of electricity is the cost of production of the energy before transmission to the final consumer.
12. These low estimates are largely due to capacity factors set out at nearly 100 per cent.
13. Below, in Chapter 6, we will see the prospects for mariculture, which is in effect farming marine flora and fauna. The first development of mariculture

and OTEC (particularly land-based OTEC) could have an impact in reducing the overall cost of OTEC electricity. The cold water pumped up from the ocean depths is very rich in nutrients and carries few diseases or parasites. It provides an excellent medium in which to culture marine algae. From this algae-rich mixture, a variety of food chains could be begun, ultimately producing either animal feed or high-value 'crops' for human consumption. One, albeit optimistic estimate, reported by the Office of Technology Assessment in America suggests that a 100 MWe plant could yield mariculture by-products of value equal to that of the electricity generated.

14. The rather 'gung-ho' tone of some political advocacy of OTEC in the United States can be understood in relation to the American space programme. There can be little doubt that OTEC was seen by many as the prestigious large-scale national project that would take over from manned lunar exploration as the showpiece of American technical achievement.

6 Energy from the sea: other proposals

INTRODUCTION

It should be clearly understood that ocean thermal energy conversion by no means exhausts the possibilities for extracting energy from the seas. Although OTEC claimed the centre of the research stage, particularly in America in the aftermath of the oil crisis, the world's oceans have long been seen as enormous reservoirs of renewable energy and many proposals have been made over the years for the extraction of that energy. The energy available in the oceans manifests itself in several ways. Firstly, and most obviously, there is the physical movement of bodies of water. This may be observed as tides, waves or currents, but it is movement that can provide the motive power for a generator appropriate for converting that movement into a more useful form of energy. Secondly, within a body of water there are gradients and from these differences within the oceans there exist potential energy differences which can be exploited to produce useful power. OTEC exploits thermal gradients but, theoretically, huge quantities of energy are also available from the salinity gradients and density gradients in the ocean. Nevertheless, the physical means of retrieving this energy will be very large and, even in comparison with the OTEC programme, very expensive. The third way in which the energy of the oceans is manifested is in the organisms of the sea. Although these are not strictly a renewable resource, they are in practice very nearly so and it has been proposed that the high efficiency of marine food chains represents a feasible way of converting the solar energy that falls onto the ocean surface into useable power. A marine plant such as kelp could be grown at sea as a crop and its biomass would be converted into a chemical fuel.

Apart from these examples of ways in which the energy within the oceans might be extracted, there are other marine energy proposals which, although not restricted to maritime operation, might benefit from being sited at sea or might be appropriate for operation in conjunction with other marine energy technologies. An example of the former is the proposal to harness the strength and also the steadiness of the ocean winds by siting wind power arrays offshore. An example of the latter is the proposal to use electricity generated by any marine energy technology to produce hydrogen from seawater by siting the hydrogen production facility at sea together with the energy production.

Other energy technologies, in themselves in no way related to the oceans, may be advanced by virtue of developments in marine technology. The improvements in the operation and design of artificial islands and other structures at sea may permit the coal reserves that lie under the continental shelves to be mined. Technical developments within the context of conventional coalmining are unlikely to enable this to happen, but the input of independent developments for marine structures may be crucial in rendering it possible to mine below the sea-bed far from shore.

In the remainder of this chapter a variety of these energy options, including proposals to exploit the energy within the oceans and proposals concerned with energy technologies sited at sea, will be surveyed. This is an important distinction because the tendency to think of ocean energy solely in terms of the former category will be seen to lead to the exclusion of several interesting and potentially very important options.

TIDAL POWER

Harnessing the tides to generate electricity is not an exotic technology; indeed it is a technology of the present, since two tidal power stations are already in operation. The Soviet Union has run a small pilot plant for over a decade near Murmansk, but better-known is the installation on the Rance estuary in France. Begun in the mid-1960s, this 240 MW generator has been technically very successful. It is, however, indicative that, even with the benefit of this experience, the French have not been enthusiastic about further developments.

The principle of tidal power is simple. The flow of the tide is channelled through turbines which turn the generators. A dam, usually called a barrage in the context of tidal power schemes, is constructed to house the turbines and to create a pound with which to regulate the flow of water through the turbines. The barrage is of conventional shallow water dam construction based upon steel or concrete *caissons*. In most tidal power proposals the intention is to allow the incoming tide to flow unrestricted through open sluices into the pound behind the barrage with no turning of the turbines. Once high tide is reached the water begins to ebb, the water in the pound can be released back onto the seaward side of the barrage through the turbines to generate electricity.[1]

The scale of civil engineering required for a tidal power scheme is considerable. The undertaking for the barrage is the major item of building; but the most serious constraint on the development of tidal power schemes is one of location. For viable electricity generation, there needs to be a tidal range between high and low tide of at least 7 m. Very few places in the world can meet this requirement. Normally it will be found in an estuary where the water gathered at a wide estuarine mouth is funnelled by the narrowing water course, thus creating large tidal changes. Not all estuaries are suitable. Apart from possessing the necessary geographical configuration, an estuary on which a tidal power station is to be sited must also have a sufficiently large volume of water flowing into it at each tide. Only a very few sites in the world offer the combination of high tidal range and very large volume of water. The lack of sites and the scale of development necessary militate against tidal power becoming significant in the next decade. It is not impossible that no new projects will be started in that time. Certainly the last ten years have seen little serious interest in tidal power in the West. Feasibility studies for the Severn Estuary in England and the Bay of Fundy in North America have been completed,[2] but the expense in relation to the current climate of energy demand does not seem to justify such large-scale projects. However, as with other marine technologies, the poor prospects for the largest schemes do not preclude smaller-scale programmes in specific locations in which tidal power may be appropriate and viable. To this end some developing nations are including tidal power among their options for indigenous energy supplies.

Small-scale tidal barrage schemes are not confined to developing countries, however. A recent study has assessed the feasibility of constructing and operating a barrage on the River Mersey in England.[3] It proposed that there were several sites on the Mersey Estuary where a power-generating barrage could be located. Depending on the precise siting, the barrage would be between one and two kilometres in length and would produce between about 350 and 600 MW. Although a major undertaking, compared to the Severn barrage proposal, it is smaller and not expensive. Power could be generated from the Mersey barrage within as little as seven years after work had begun, which illustrates very clearly that it is a project on a much more easily managed scale than the Severn scheme would be. The Mersey barrage may never be built but the studies for it show that the Bay of Fundy and the Severn Estuary do not exhaust world options. Indeed, one of the points made in the Mersey discussion was that such programmes might be justified, in part, as pilot plants for larger schemes, quite apart from any inherent advantages of their own.

Wherever barrages are built there will be benefits over and above the generation of electricity. Behind the barrage an artificial lake is created, which would not only be available for leisure activities, but might also provide an opportunity to develop deep water harbour facilities on what was previously a tidal estuary. In parts of the world where estuarine flooding is a recurrent problem the barrage and the pound could be used to control the flow of water into the estuary and so reduce the risk of flooding.[4]

Despite twenty years of experience of them, there do remain some uncertainties as to the long-term impact of barrages, which might have adverse consequences for the construction of tidal power schemes. These uncertainties relate to the changes of tidal flow in estuaries after a barrage has been built. The immediate effect of the barrage is to raise the mean water level on the landward side of the dam, where higher water levels are retained longer than under tidal conditions because water is held in the pound after the natural tide has begun to ebb. Also, the lowest water levels in the pound are raised above the natural low-tide level. So, not only is the average water level in the pound raised, but the tidal range between high and low tides in the pound is also reduced to about half of the natural tidal variation. This has obvious ecological consequences. Habitats such as mudflats will be submerged for longer, if not permanently flooded. But the retention of extra water behind the barrage has other consequences which could ultimately threaten the long-term viability of the schemes, owing to the changed characteristics of water-flow in the estuary. There might be silting up in the channels on both sides of the barrage. This would be counterproductive to any prospects of developing harbour facilities behind the barrage but, more seriously,[5] in the long term siltation could sufficiently reduce the flow of water into and out of the estuary to render the generation of electricity uneconomic. This prospect of self-destruction weighs heavily against tidal schemes. The long-term consequences of the barrages are not fully understood and this rather suppresses enthusiasm for the adoption of large-scale tidal power schemes in the near future.

WAVE POWER

The generation of electricity from the motion of the waves has received plenty of attention in the past. The continual breaking of waves against the seashore and the damage that waves can wreak serve to demonstrate the energy resources of the ocean swell. Geographically, the locations to which wave energy schemes would be applicable are as specific as for OTEC or tidal power schemes. The requirement is for a long coastline with open sea beyond, off which the ocean winds and waves can sweep uninterrupted. These conditions are normally found in the middle latitudes on the margins of the great oceans and often on the western coasts of continents.

Many designs for the extraction of wave energy have been proposed.[6] The early designs, and many of the recent ones as well, were usually based on the motion of a floating buoy which rises and falls with the waves. The energy of the waves' vertical motion is hence transmitted to the buoy and the mechanical energy of the moving buoy turns the generator to convert mechanical energy into electricity. The design of the floats in recent schemes has become very sophisticated in order to remove as much energy as possible from the incoming wave. In the most efficient designs the extraction of energy is so complete that the water behind the buoy has hardly any wave motion on it at all. Predicated upon the operation of this family of designs, proposals for large-scale wave power arrays have been offered.[7] Thus far, nothing has transpired.

The biggest single drawback to the float system described above is that it is an electro-mechanical device by which wave motion is transferred mechanically to a generator which then converts the mechanical energy into electrical energy. The theoretical efficiency of this chain of conversion is held back at each stage by the thermodynamic limits and, despite sophisticated recent designs, the frictional losses in the system are still significant compared to the relatively low potential energy of each individual wave. These two problems limit the overall efficiency of electro-mechanical conversion of wave energy and had much to do with the loss of interest in simple wave power devices in the face of the United States' commitment to OTEC as the primary hope for marine energy development in the 1970s.

There are two principal ways of solving the problem of wave power conversion efficiencies. Either the potential energy of the individual wave must be amplified or an alternative to electro-mechanical power generation must be found. The former can be achieved by a number of designs, several of which received attention in the 1970s. One solution to wave amplification is the 'Dam Atoll' proposed and pursued by Lockheed in the United States. It consisted of a large floating concrete dome, some 75 m. in diameter, which floated with its top just level with the sea's surface. At the crown of the dome there would be a hole through which the waves would enter the structure. The water spirals down into the core of the unit and turns the generator turbines directly. The design benefits from having very few moving parts and the chain of energy conversion is shorter since the water itself falls onto the turbine blades, with no intermediate mechanical phase. The Dam Atoll concept was in keeping with the fashion of the time in the United States of conceiving of marine energy projects on a very large scale. Each dome would have an output of between 1 and 2 MW and the output from a full array would be sufficient to make a significant impact in terms of growing national energy demands. This concept of marine energy was ideal for the high-technology corporations such as Lockheed which sought the federally financed research contracts on alternative and renewable energy schemes. Dam Atoll was an expensive, large-scale engineering project which, in the event, lost out to OTEC as the vanguard of America's ocean energy programme and which, having failed to gain that commitment in the 1970s, seems to have few prospects of commercial revival in the immediate future.

Rather more practical for the near future than Dam Atoll are those designs whereby the waves' motion is amplified by their impact on a column of air. The principle was field-tested off Hawaii in 1976 and mechanical energy was produced by a 300 ft long 'pump' which effectively magnified the wave pressure head. At the time (when it was important to talk of large systems) it was claimed that, with a 1,000 ft long pump, up to 1 MW of power could be produced. Also, since the mid-1970s the Japanese have pursued their own version of wave power amplification. The Japanese Marine Science and Technology Centre, formed in 1971, has been conducting wave power research since its early days. Recently it has concentrated on the *kaimei* system of amplification in which the vertical movement of the seawater is transmitted into the motion of air within a floating barge. Sea trials were run in the late 1970s, some of which were undertaken in co-operation with the United States, Canada, the United Kingdom and Eire.[8] The test barge of 800 tonnes carried eight generating units of 125 kW each, and in 1980 electricity was transmitted to shore from the barge. Further phases of research and development have been pursued through the 1980s, with the principal aim of testing turbines that can operate most efficiently at the relatively low energies produced by the wave head even after amplification. The Japanese aim to complete tests in the late 1980s and then, in the manner favoured by the Japanese for new schemes, private enterprise will be expected to take the project on to commercialization if it has proved technically successful.

Both these methods of wave amplification are still fundamentally electro-mechanical in that the aim of amplification is to produce a larger wave pressure head to drive the turbine. Such methods, however sophisticated, will suffer losses in the efficiency of conversion. Hence, there are proposals for an alternative to electro-mechanical systems. This option began to receive attention in the late 1970s and early 1980s as new polymers were being developed. Some of these materials possess the property of producing an electric current when subject to pressure changes. Using such polymers, the pressure changes of the wave head could be used to generate electricity directly, without the intervention of other mechanical systems. Amplification of the wave head could be used in conjunction with electro-chemical methods of power generation to further enhance the efficiency of the process.

In general, wave power has an appeal, but it becomes difficult to operate in practice since, apart from the Dam Atoll scheme, wave power is not ideal for the generation of very large amounts of electricity. If a conventional wave power array is to generate a large power output, it requires the harnessing of a very long wave front. In engineering terms the problem is to maintain the integrity of the array along its length since, by its very nature, a wave power scheme will not be based upon a solid and compact structure. Yet it must also be able to absorb the full force of the ocean waves. The second major problem of wave power is inherent in the efficiency with which the designs can extract the energy of the incoming wave front. Behind a large array there would be a lagoon of calm water, in which there would be little or no wave action. Without the action of waves, the environment of the sea behind the array will be changed and this will be especially severe if the array is in shallow water or close to shore. Patterns of sand deposit and scouring would be altered, which would have an ecological impact but would also have potentially damaging consequences for the commercial activities of the coast behind the array if shipping channels and harbour facilities were disrupted by those changes.

OCEAN CURRENTS

The energy of the currents of the oceans is the least accessible manifestation of the power of moving bodies of water. Yet there have still been proposals for harnessing the energy of those movements. The Gulf Stream in the North Atlantic, suggests one proposal, could be tapped using an array of 242 large turbines sited in the current over an area of 60 by 30 km. off the coast of Florida. Each turbine would be 170 m. in diameter and 100 m. in length, with rotor blades 90 m. in diameter and an output of 83 MW. In its scale it is almost impossible to visualize and it was one of the most exotic of the schemes for which the United States government financed feasibility studies. Taken by itself, it is not without advantages—the environmental impact would be minimal and its supporters have claimed that it would be economically competitive with other ocean energy proposals.

The Japanese have also looked seriously at the energy options for the Kuroshio current, which is the Gulf Stream's equivalent in the North Pacific. But harnessing the energy of either the Kuroshio or the Gulf Stream requires expenditure and development on a scale that is not appropriate to energy needs as foreseen for the next two or three decades. Even when ocean energy systems become technically and economically attractive for large-scale programmes, it is likely to be OTEC and wave power schemes that will hold greater promise than those of the ocean currents and the option of power from the currents will not be an early choice for development.

SALINITY GRADIENTS

Fresh water can be retrieved from brine by a number of methods, but all require a net input of energy. Therefore, when the reverse process takes place, energy is liberated. This is the principle of salinity gradients as a source of power in which bodies of water of differing salinity can be mixed to liberate energy. The greater the salinity gradient, the greater the energy available and so sites at the mouth of rivers would be able to utilize the gradient between seawater and the freshwater outflow of the river. The theoretical amounts of energy available are very large.[9] There are several conceptual designs which would permit the energy of salinity gradients to be tapped. The two principal groups of designs are based upon simple reversals of the desalination process. In one design the waters of differing salinity are separated by a semi-permeable membrane. By osmosis the solutions will mix through the membrane to equalize the salt concentration between the two. The movement to equilibrium will be exothermic and the liberation of energy will be manifested in a temperature rise of the mixed solutions. In the other system the process is the reverse of the electrolysis that is used for desalination, in which two solutions of differing salinity are separated by an electrically conducting membrane and connected to an electrical circuit. A voltage is produced in the circuit from the ion transfer between the solutions. On a large scale, electrodialysis could be used to generate electricity directly from saline solutions. Both osmosis and electrodialysis are simple reversals of energy-intensive desalination processes. The principles of both have been demonstrated and patents granted on detailed designs, but neither can be said to be seriously practical even in the most specific circumstances. Commercial schemes would need to be large and would involve high capital expenditure. It is not likely that this would be worthwhile for any energy resource which, although huge in total (since solar energy will always provide the cycle of fresh water from the rivers into the seas), is thermodynamically low-grade.

DENSITY GRADIENTS

Similarly to the salinity gradients between bodies of water, there are also density differences. Such gradients are normally caused by temperature differences,[10] either from the cold water of the ice flows melting in warmer seas or from hot springs producing localized temperature differences compared to the surrounding body of water. But it is from the density gradients consequent upon these temperature differences rather than the temperature differences themselves (as in OTEC) that energy could theoretically be extracted. Although its proponents claim that density gradients represent a higher grade of energy than either thermal or salinity gradients, much the same conclusions can be drawn for this proposal as for salinity gradients and ocean currents as a source of significant energy production. The scale of the programmes needed to utilize these energy sources has to be so large in order to be feasible and there are other energy technologies, including some of the marine programmes, which show promise of being developed much more easily, and initially on a smaller scale, than these abstract proposals.

BIOMASS ENERGY

Even within an activity so ripe with the exotic as marine technology, the notion of growing fuel in the sea may sound unlikely. It is nevertheless a serious proposal and rather more practical than some of the others that have been discussed. It exploits the high efficiency of marine food chains to produce biomass from the nutrients in seawater. The crop could be one of several species ranging from micro-organisms such as algae through to large marine plants like kelp. The advantages of farming marine biomass at sea are the absence of competition for the use of the site (as there would be on land) and the rapid growth of the crop on abundant and cheap nutrients.

The biomass, once it has been harvested, can be converted into fuel. Simple combustion of the biomass would be inefficient and uneconomic, instead it is converted chemically into dilute organic acids and hence to olefins and hydrocarbons, mainly methane. The chemical engineering does not present serious problems and several chemical routes for conversion are available. Anaerobic digestion of the biomass to yield methane and carbon dioxide has some advantages, not least that it leaves the nutrients, in particular the nigrogen, in the residue which can subsequently be used as fertilizer.

The capital cost of a hypothetical biomass farm might be lower than expected since it could be run in conjunction with, say, an OTEC plant, which would already be pumping deep ocean water to the surface. This water is very rich in nutrients and relatively free of diseases, thus providing an ideal medium for marine vegetation to achieve rapid growth rates. Since the artificial welling-up of deep water would be an integral part of an OTEC plant, marine biomass growth could benefit from the upwelled water without the capital cost being charged against the biomass programme. During the 1970s, extensive research was carried out in America into the growth rates and yields of various marine organisms in order to determine the optimum conditions for their culture, including the influence of upwelled water. In Japan a small pilot study on kelp was run in the late 1970s, and the 1980s have seen a more extensive programme to assess the value of artificial upwelling to mariculture projects. These studies are fundamental to the prospects of biomass energy production, but there has been little interest so far in proceeding beyond these initial studies except on paper. Obviously no large-scale impact on energy supply should be anticipated from biomass energy; it is essentially a small-scale concept that might be feasible in conjunction with other schemes rather than in its own right. Nevertheless, it does offer one important advantage in that it provides its energy in chemical form. Not only is this more appropriate for certain uses, but it is also much easier to transport it from an offshore site than it is to transmit electricity by underwater cables which present a severe and critical problem for all the marine energy technologies which directly generate electrical power.

HYDROGEN PRODUCTION AT SEA

The technical problems arising from the transmission of electricity from offshore production sites to land can be avoided if another technology is employed at sea in conjunction with the electricity generating system. The electrical power generated directly from, for example, an OTEC plant could be used in the electrolysis of seawater in order to split the water molecules and evolve hydrogen gas. The hydrogen, once liquefied on site, can easily be transported and stored. The principal requirement for hydrogen production at sea is the provision of the electrolysis and liquefication plant on site. This will require the development of artificial islands—probably floating platforms for deep water operation—capable of carrying medium-sized chemical engineering installations. In Chapter 7 artificial islands will be discussed in detail, but for the present suffice it to say that the construction and operation of industrial plants on floating artificial islands appears to pose rather less daunting technical challenges than those of some of the other novel marine technologies. If electricity generation at sea were achieved, it is unlikely that, if desired, a hydrogen production facility could not be included within the complex since the energy production system itself might well involve artificial island development quite sufficient to meet the needs of a hydrogen production plant.

OCEAN WIND ENERGY

As with proposals for wind power on land, the concept of generating electricity from the winds over the sea has received steady though hardly intense interest. Possibly the image of wind power on land implies a technology that is only appropriate on a very small scale, but this is not necessarily the case for ocean wind power. Certainly it would be misleading to think of ocean wind power in terms of floating windmills! The technology is simple compared to other marine energy proposals but the designs—as indeed they now are for land-sited windmills—are for sophisticated rotor generators with very high efficiencies. When ocean wind power has been considered it has not usually been considered in isolation, but rather as an adjunct to wave power. Since those locations favoured by uninterrupted sweeps of waves from the open oceans will normally also receive steady strong winds off the ocean, it has been suggested that the two systems should be sited together as a combined wind/wave array. There would be economies of scale in construction and an element of 'belt and braces' in exploiting the same basic energy resource in two different ways.

Wind power at sea should not be confused with the single pylon rotors used for water pumping or electricity generation on land. On land they are used for very specific local needs—perhaps the supply of electricity to a single farmhouse. At sea it is possible to have an array of generators whose generating capacity, if not large in terms of national grid requirements, can be significant to the needs of immediate coastal communities. However, whether as part of a combined wind/wave system or as ocean wind power by itself, although relatively cheap and simple, it is unlikely to be adopted widely in the medium term. Nevertheless, development may well continue and if wave power, which does have better prospects, is adopted as a viable energy-generating system, then ocean wind power may have a future as part of that development.

UNDERSEA COALMINING

The last of the energy technologies to be considered here is not primarily a marine technology at all, but if it is to be developed then advances in other specifically offshore technologies will be necessary. For centuries, coal has been mined from seams running out from land under the sea-bed. But this mining from pits on shore can only extend a few kilometres out to sea, whereas the coal measures may extend very much farther under the continental shelf. The coal measures under the sea-bed are estimated to be very large indeed and the extension of British coalfields under the North Sea is certainly one area in which the future possibilities for mining coal under the sea-floor is creating interest, since drilling samples have shown there to be deposits of quality, high-carbon content coal. Since mining these measures from pitheads on land is impossible[11] there is obviously a need to site the pithead at sea, either on the surface or on the sea-bed, and it is here that the project requires the assistance of new marine technologies. Developments in marine structures might permit genuinely independent offshore mining to begin. Full pithead installations would need to be incorporated in a platform—most likely a concrete gravity platform—which would be floated to the site, anchored and then shafts sunk prior to mining. But even this option would not be appropriate for the deepest coal measures far from shore. Conventional mining, however, automated, will be limited by depth and the heat within the deepest measures. In the longer term, therefore the exploitation of the bulk of the coal reserves below the sea-bed requires a different approach. That approach is probably *in situ* gasification of the coal, by which the coal is converted, while still underground, into gas which can then be extracted from the coalfield by techniques similar to those employed on natural gas wells. The gas is easily transported either by pipeline or by tanker, perhaps after liquefication on site.

Several coal gasification processes are available. They have been demonstrated and some are running commercially, although in most cases the processes are too expensive for extensive commercial operations.[12] The end-products of gasification are hydrogen, methane, carbon monoxide, carbon dioxide and water. The efficiency of the process relies upon achieving a reaction which produces the most methane. The process is not chemically complex, but it requires the coal to be in fairly small pieces and raised to temperatures in excess of 500°C in the presence of steam. This presents no severe problem for processing on land, but it may not be a trivial problem to achieve those conditions underground and then subsequently to control the reaction.[13] If *in situ* gasification were to become a reality, its application to the extraction of the coal from undersea measures would be dependent upon deep sea structures from which the initial drilling could be undertaken, the process initiated and controlled and the gas products extracted prior to transporting away from the site. At present the availability of coal that can be won by conventional means renders undersea *in situ* gasification unnecessary, but if that option were pursued in the future it would require as much in the development of marine structures as it would in the extractive technology itself.

COMMENTARY

The variety of the ocean energy proposals reveiwed above aptly demonstrates the creative effort that has been put to the service of harnessing the oceans' huge energy resources. The fact that the energy manifests itself in so many different ways only gives rise to a plethora of schemes, each seeking to harness one particular form of ocean energy. It provides an interesting comparison with the projects for extracting minerals from the sea. These proposals seem very prosaic by the side of the energy schemes, but a mineral is either present or it is not and this fact rather limits the scope for imaginative proposals. This is not so with ocean energy where each manifestation of the energy could theoretically provide a source of power, and a design, no matter how exotic it might seem, can usually be conceived to harness that energy. This makes it difficult to assess the serious prospects for ocean energy technologies. Most of the proposals are exotic, most are highly capital-intensive and most seek to exploit energy in forms that, though large in size, are thermodynamically of low grade. Furthermore, many of them are the products of the late 1960s and early 1970s, when the implications of extrapolated energy demand growth rates and the energy price rises subsequent to the oil crisis provided an impetus to a whole range of alternative, and often renewable, energy technologies.

With hindsight, some of them appear so fanciful that it is difficult to give them any credibility in view of the energy economics that have prevailed in the Western economies since the late 1970s. But they had to be taken seriously at the time because conventional energy supplies appeared limited and vulnerable. It should not be forgotten that OTEC is as exotic as any of the proposals and yet it was OTEC that, with political and financial support in the United States, made such strides during the 1970s. With that level of commitment, almost any of these marine technologies could be brought to pilot plant operation and so it is unwise to dismiss any of them out of hand. Certainly, in the early 1970s, the Dam Atol wave power proposal gained considerable support and if it, rather than OTEC, had received political and financial commitments from the Carter administration in America, then it might be Dam Atol plants which were now nearest to commercial operation.

Nevertheless, circumstances have now changed. The context of the economic and political future of energy provision is very different from what it was in the early 1970s. The rate of growth of energy demand has ceased to increase so rapidly and the continued glut of oil on the world market has reduced the threat of oil being used again as a political weapon. This has tended to reduce the scale upon which alternative energy programmes are conceived. Novel marine energy schemes are no longer seen in terms of their potential contribution to national energy consumption, rather they are now seen, at least initially, as small-scale technologies which may be appropriate for development in specific locations with specific requirements. Even OTEC is now being pursued in this way. Any of the energy technologies we have looked at may have a future, given a suitable constellation of factors (political and environmental as much as economic and technical) on a small scale for an appropriate situation. Some of the schemes might seem to have better chances of development than others, in so far as some are more suitable for small-scale operations, but any future fears about the supply of energy from conventional sources could bring any of these technologies, even the largest and most exotic ones, back into very serious consideration.

NOTES

1. The Rance scheme does in fact generate electricity on both the ebb and flow tides. This is not normally proposed since there are design complications if the system is to operate on both phases of the tide.
2. These are the two locations most often cited as suitable for large-scale tidal barrage construction. The Severn Estuary proposal has rumbled on in Britain for many years, leading to the completion of some substantial reports in the early 1980s.

3. In the early 1980s pre-feasibility studies for a Mersey Barrage were commissioned by Merseyside County Council. Much of this work was undertaken by a research consortium comprised of groups from universities in the North West and called Marinetech North West.

   In late 1986 the Government and the C.E.G.B. signed contracts worth £500,000 to continue and extend the studies that Merseyside had been financing.

4. This is the way in which the Thames Estuary flood barrage operates, without producing electricity, by controlling the inward flow of tidal water and the downstream flow of water from the river.

5. Apart from the retention of silt behind the barrage, the reduced flow of water may lead to a build-up of pollutants in the pound. If this were the case the prospects for leisure use of the pound would be reduced.

6. Summaries of patents in the field of marine technology in the 1970s clearly show the large number of patents taken out, by individuals as well as firms, for wave energy devices.

7. The best-known of the floating buoy designs are those following from the work of Salter, commonly known as 'Salter's Ducks'.

8. These are all countries, it should be noted, which have coasts on the western sides of continental land masses.

9. The large quantities of energy available are clearly demonstrated by the converse fact that the desalination of water is prohibitively expensive in terms of energy except for the most special cases.

10. Differences in salinity would also give rise to slight density differences.

11. Mining from land can be extended out to perhaps ten kilometres from shore by sinking satellite shafts from concrete gravity platforms. These shafts would carry electricity and provide ventilation, but the main access to the coal faces would still be on land. Thus, this option does not permit mining beyond the margins of the oceans.

12. For mainly strategic reasons, South Africa has commercially pursued several methods for producing synthetic oil and gas from coal.

13. Analogous techniques, including pulverization, have been proposed for the *in situ* extraction of oil from shales and oil-bearing sands. Thus there may be an input of experience from a related technology to coal gasification.

7 Artificial Islands

INTRODUCTION

The notion of new land created on the sea or from reclaimed land is not novel, but now a wide range of projects are being assessed for a whole variety of specific industrial uses. 'Artificial islands', which serves as the generic term for such constructions, includes *polder* reclamation in Holland, many floating structures used in harbour installations and most of the structures developed for the offshore oil and gas industry. The techniques which have served in these existing fields are largely suitable for future developments of larger artificial islands and for their utilization in other ways. These new functions may include services in support of marine energy and mineral projects discussed elsewhere in this book.

Artificial islands are attractive for a number of reasons. A range of designs are available to suit particular requirements in specific locations, but the advantages of artificial islands can be summarized under general headings. Firstly, artificial islands can create new land where either no land exists or where it would be prohibitively expensive. Artificial islands might permit expansion of an industry near to its markets, near to its raw materials, or near to related industries. An artificial island allows a new site to be developed from scratch without the constraints of fitting into an existing location. Without artificial islands, none of the developments would be possible if land on shore is scarce or expensive. The second general advantage is that artificial islands can be used to remove industrial processes that are dirty or potentially hazardous away from human habitation to more remote areas. Airports, with their attendant noise pollution, might be built offshore, while waste disposal and even, possibly, nuclear power stations might be more acceptable if isolated away from land. The third category includes those cases where artificial islands enable an industrial plant to be sited somewhere that is inacessible or inappropriate for conventional construction. One example of this would be the provision of complete factories for developing countries, where the facility can be floated in all ready for operation. A fourth area where artificial islands could prove valuable is in relation to activities that take place by, or could usefully be placed by, water. Obviously, the construction of deep water harbour installations or fish processing plants assumes the proximity of water and could benefit from artificial islands. Power stations need water for cooling, for that reason, they could be built on artificial islands.

Similarly, water is important for many leisure activities and facilities for these could be sited offshore rather than onshore. The final category in which artificial islands would serve is in association with the development of other marine resources. Currently, the best example would be in relation to offshore oil and gas production, but many of the other projects (such as OTEC, manganese nodule mining or phosphorite mining) that might be of future importance all rely on the viability of operating at sea. The development of artificial islands appropriate to the specific requirements of each project may be crucial to ensuring success.

DESIGN AND CONSTRUCTION OF ARTIFICIAL ISLANDS

There are four families of artificial island designs. Which type is chosen for a particular purpose will depend, in part, on that purpose and, more importantly, on the characteristics of the site. The features of a location that would influence design choice include depth of water, tidal range, geology of the seabed, risks of damage and the proximity of shipping lanes or fishing grounds.

The first family of designs are the floating islands. These consist of a raft, usually of concrete or steel, with the operational facility placed on a platform on top of the raft. This mode of construction renders it possible to prefabricate rafts of specified dimensions from standardized components. Prefabrication offers reduced costs and also the opportunity to build rafts for smaller projects as well as the largest ones. Floating islands have the advantage of mobility and can be moved either for maintenance or for operations at a new location as required. In deep water, where the sea-floor is unsuitable for other types of artificial island, or where tidal changes are very great, floating islands are an appropriate and flexible option.[1] The main disadvantage is vulnerability to damage from collisions and even the chance of sinking, especially in severe weather. Also, in adverse weather floating islands may lack stability, which might interfere with the performance of operations on board and might also render access for personnel and materials difficult. Constructionally, there are no grave problems to hinder the increased use of floating islands, particularly if prefabrication of standard units is pursued, and a variety of concrete and steel floating structures have been used for many years in a range of maritime operations. The one serious technical problem may be the provision of moorings for very large structures in deep water. Stresses on cables and their fixings will be very high and there must be no possibility of a large deep ocean platform breaking free of its moorings even in the severest weather.

The second group of designs are the fixed islands, which are structures with legs attached to the sea-floor supporting a platform above sea level. Typical of these are the rigs that have been used by the offshore oil and gas exploration industry where two types of fixed islands have dominated. The gravity platform is a concrete structure resting directly on the sea-floor and supporting the platform, usually on a single central column. The alternative is the steel-legged platform. This has been the more common choice for oil rig design up to the early 1980s, but concrete has become increasingly popular, particularly as oil exploration has moved into deeper water. Many of the smaller artificial islands programmes borrow directly from either the concrete or steel fixed islands developed for offshore oil. For larger and more ambitious projects a third type of fixed island exists. In this case steel or concrete *caissons* are floated to the site and then sunk. The platform is then constructed on top of the *caissons*. A refinement of this would be to sink piles into the sea-floor and then add a concrete deck. *Caissons*-mounted fixed islands are based on practices already used in harbour construction and therefore all the fixed island designs (that is including steel-legged and concrete gravity platforms) draw upon tried and tested techniques. It is in the extension of these techniques rather than in the development of novelty that the future of this group of artificial islands seems to lie. As a group, their great advantage over floating platforms is increased stability and resistance to severe weather. However, they are still vulnerable to collisions and the problem of access in bad weather is not entirely eliminated for fixed islands. In use they are restricted to the shallower waters and to areas not subject to the widest tidal ranges,[2] nevertheless, within these limits (and at sites also providing a suitable sea-floor for anchoring the fixed elements) the advantages of prefabrication and the depth of existing experience in construction make fixed islands an attractive option.

Whereas fixed islands consist of a platform held above water by some suitable structure, the third family of designs represents more truly the creation of new land. These are the fill islands. Using the techniques of *polder* reclamation, fill islands genuinely increase available land rather than merely providing a steel or concrete platform in the sea. The simplest fill island is created by dumping sand or gravel at the site and building up the level of the sea-bed until it is raised above sea level. The dumped material settles to form a natural slope at the edges which, in effect, provides beaches surrounding the island. As with natural beaches, these do much to resist the action of waves, tides and currents and so give protection to the interior area. The beaches can be reinforced and the slopes of the island con-solidated to give extra resistance to erosion.

An alternative is to surround the area to be reclaimed with an embank-ment.[3] The enclosed area can then be pumped dry. Pumping can continue and the floor level can remain below sea level. Although this is expensive, it can be appropriate for large areas where the costs of filling and constructing a platform might be more than the cost of pumping. In other cases, of course, where filling material is available cheaply, the *polder* could be filled to a height above sea level to avoid the need for continual pumping. Whether the island is filled or left as a *polder*, the embankments must be able to resist effect of erosion by the sea; but so long as this is achieved fill islands are extremely stable and not at all vulnerable to damage. The main advantage of them is that very large areas can be made available by relatively simple construction techniques. Obviously, of course, fill islands are restricted to shallow water sites and, once built, their impressive stability means it is unlikely they could be dismantled. Compared to floating and fixed artificial islands, fill islands are permanent, lacking in flexibility and suitable for a more restricted range of locations; however, fill islands do offer size and stability beyond that available from others, providing that the beaches and embankments retain their integrity. The care of these margins is the only maintenance that fill islands require and running costs once construction has been completed are low.

Floating, fixed and fill islands all have their respective advantages. The fourth and final group of artificial islands are the composite islands, which seek to combine the designs for the benefit of specific uses. For example, a floating island could be protected from severe weather in the open sea if it were moored within an artificial breakwater (either fixed or floating itself). Within the artificial lagoon, the floating island is protected from collisions as well as heavy seas, but the advantages of the floating island could still be utilized in construction and operation. Another scheme appropriate in certain circumstances would be to build a fixed island with the operational facilities in place on its platform. The island, possibly on a *caisson*, would be towed to its location and fixed in place. A fill island could then be constructed around it. In this way the plant itself can be assembled onshore rather than needing to be undertaken at the operating site, while the subsequent provision of a fill island to surround the plant offers the stability and security, as well as the maintenance advantages, of that design. This is an option for shallow water operations, whereas the artificial lagoon around a floating island would tend to be deployed for deeper sites; but other combinations can be devised with benefits appropriate to particular purposes.

For almost all artificial island designs there are areas where a choice must be made between concrete and steel in construction. For many uses either would be adequate, although the intended use and the anticipated life required of the structures do, in some instances, favour one rather than the other. In the North Sea both have been in service but the design lifetimes for oil exploration are normally only about twenty years. However, artificial islands may be required to operate for considerably longer than twenty years, especially for uses such as waste disposal or waste storage. In view of the need for longer periods of operation, it appears that for elements which are below the surface of the sea concrete will be most suitable. Concrete has been used in a variety of marine environments for many years and the experience gained permits confidence in the assessment of design and performance. For underwater applications concrete is less prone to corrosion and easier to maintain than steel would be in the marine environment. However, above water level, on the platforms, steel may be preferred. Construction is usually easier with steel and in most cases the total weight of steel required for a particular job is less than the weight of concrete would be for the same purpose. Notwithstanding its tendency to corrode, routine maintenance is easier for steel than for concrete and for platforms and plant facilities above the water level steel will often be chosen. Although new materials may be useful, they are not essential for artificial islands to be built. The designs described above could all be constructed using conventional materials—primarily steel and concrete.

EXAMPLES OF ARTIFICIAL ISLAND PROJECTS—1: BARGE-MOUNTED PLANTS

The requirements of a number of industrial processes suggest that some might be suitable for mounting on floating platforms. One project is for the liquefication of the natural gas found associated with offshore oil reservoirs. Crude oil under pressure in underground reservoirs contains light hydrocarbon gases. Once the oil is recovered and is no longer subject to high pressure the gases present a problem. It is not particularly easy or economically sensible to pipe the gas to shore or to transport it in tankers. At land sites the gas is processed to recover propane and butane which can be liquefied for ease of transportation. The recovery is worthwhile since the gases constitute about 15 per cent of the energy value of the crude. At offshore production platforms, processing the gas has not so far been feasible and so the practice of 'flaring-off' has been the only option. The only option, that is, unless a processing plant on an artificial island made liquefication at sea viable. It would almost certainly be a floating platform, which could be built complete on shore and then towed to the operational location. An additional advantage is that a floating platform could be moved between oil fields as required, hence gas liquefication could be undertaken even for small oilfields in deep water where a purpose-built fixed platform would be too expensive.

Once constructed, the floating plant obviates the need to use facilities on shore. Not only is a land process plant unnecessary, but so also are deep water harbour facilities for liquid gas tankers as all the docking and loading will be carried out at sea. Furthermore, the need for that most vulnerable and expensive adjunct to offshore oil and gas production, the underwater pipe line, is greatly reduced. However, the proposal is not entirely without drawbacks. Instead of a single process plant sited on shore, each floating plant will need to carry not only the fractionation plant and the refrigeration plant, but also a power plant, storage facilities for the liquefied butane and propane and all the ancilliary apparatus for transferring the product to tankers. This implies the operation of a number of small separate units where economies of scale may be available from the operation of a single large plant sited on land. Whether, on balance, the advantages of floating liquefication plants are sufficient to justify the project will depend very largely on the design of a compact and efficient fractionation plant. This is best achieved by low temperature fractionation and so requires the provision of refrigeration plant, which would be needed anyway, however, for the storage system (in which the product will be stored as a liquid), and so low temperature fractionation is a very convenient option.

The platform itself could be of several designs, depending on specific requirements. At shallow water sites a semi-submersible concrete pontoon supporting a raised deck could be floated onto the site and ballasted to sink to the sea-floor, leaving the deck above water. Once operations are completed, it could be refloated for towing to a new location. But, for a restricted number of shallow sites, fully floating platforms are more likely to be chosen. In sheltered sites a simple moored concrete platform carrying the plant would be adequate. For slightly more exposed conditions, a pre-stressed floating concrete structure would still be suitable, but some of the steel storage tanks should be placed below the water line to give extra stability. In the fully exposed sites, such as most North Sea locations, a semi-submersible floating platform is necessary. Storage and ballast are provided in submerged hulls, from which vertical columns support the deck with the processing equipment above water. The hulls provide stability and the platform itself is only slightly disturbed even in very severe sea conditions.

There is a school of thought that believes that the first widespread use of barge-mounted industrial plants will be in the Third World, where Western companies and Third World governments may find artificial islands convenient vehicles by which to establish industrial facilities at difficult sites or where the necessary infrastructure for construction is absent. If the industrial plant were constructed as a floating platform in a conventional shipyard it would avoid the problems inherent in building at the intended site of operation. Such sites may be isolated or be subject to unfavourable climatic conditions or they may lack a local labour-force. Any of these factors, and others too, could render conventional building wholly inappropriate in certain locations and in such places a prefabricated barge-mounted plant has considerable appeal. An example of this approach is a pulping plant built in Japan for use in Brazil. The plant is mounted on two platforms, each 225 by 45 m. and weighing 30,000 tonnes. One platform carries the pulp mill and the other the power plant. The mill was to be located at a site 400 km. upriver from the mouth of the Amazon. It was a remote site and building directly on it would have been expensive and time-consuming. By building on the rafts in Japan it was estimated that the construction period of two years was, perhaps, only half that of construction on-site in Brazil and costs were reduced by between 15 and 30 per cent. These savings were available because all the building, fitting and testing was performed at the same site as the construction facilities and the skilled labour-force. All that had to be ferried to Brazil was a complete and operational pulp mill rather than a continual stream of equipment and Japanese manpower throughout the construction period. The barges, when completed, were towed on a three-month voyage to Brazil. Once on-site, the platforms were floated on the river to their final sites by an arrangement of locks, dams and canals specially built for the purpose. The final pound was drained of water and the platforms were grounded away from the river and ready to work.

The obvious problem with these operations, despite their proven feasibility, is that construction away from the final site inevitably precludes any participation by the host country. There is no work for local labour and no opportunity, therefore, for them to learn the skills which would make other projects possible by the use of local labour. It is a vicious circle and the interest of Third World governments in such programmes is certainly therefore reduced. Moreover, the programmes run entirely contrary to current development aims which emphasize the importance of creating in the Third World the very infrastructure that would render the importation of a complete prefabricated plant quite unnecessary. Keen as Western companies may be to build or operate prefabricated floating industrial units, they will find opportunities considerably restricted in the developing countries unless the package can be shown to possess real benefits for the host country other than merely providing 'off-the-peg' Western technology.

EXAMPLES OF ARTIFICIAL ISLAND PROJECTS—2: KOBE PORT ISLAND

The Japanese are heavily committed to artificial islands for a large variety of uses. The pulp plant previously described was built in Japan, but the Japanese are equally committed to using artificial islands in their own waters. The case for artificial islands in Japan is geographic and economic. Space for industrial growth is very limited, especially near the coast, and any new land created by artificial islands will be very valuable. To this end several projects are either under construction or in the process of planning.[4] Artificial islands for primarily industrial expansion have been built in Japan since the 1960s; they have provided the Japanese with considerable experience of their construction and some notable successes. These early projects were sometimes relatively small in scale and built for a specific purpose. The industrial artificial island of Ohgishima in Tokyo Bay was built between 1971 and 1975. It was a reclamation project off-shore of the West Tokyo industrial estate on which a new steel plant had been built. The existing industrial estate could no longer provide land for expansion of the old steelworks, which needed to expand or risk becoming uneconomic. Since expansion within the existing site was impossible, a radical option was adopted to built anew on a fresh, reclaimed site away from the congestion of the old industrial estate. In this case the only way to acquire such a site was to build an artificial island.

The Japanese are now undertaking much more ambitious reclamation programmes that are not solely confined to creating new land for industry. An example of the new style project is the Kobe Port Island scheme. Kobe City is on the coast near Osaka, it has a population of four million people and covers an area of 540 square kilometres. The commercial centre is concentrated in a narrow coastal strip which includes the harbour facilities, but which has no scope for further expansion on land. So attention turned off-shore to provide space not only for extended port facilities, ship-building yards and factories for heavy industry, which were the city's staple activities, but also for commercial, residential and leisure areas. Planning began in the late 1960s and the Port Island project, as it was known, was intended to devote about a third of its area to harbour facilities, including container berths, liner berths and freight handling yards.[5] The remaining area was divided between the other uses. The real advantage of this approach was that all the functions could be integrated at the design stage, rather than as a product of piecemeal expansion, and so permit genuine rationalization of the variety of different activities.

The project was a reclamation island, a variant of the fill island, built adjacent to the coastline in shallow water. The filling material was excavated from two sites in the Rokko Range of mountains behind Kobe. (It was this range that hemmed the city into its narrow coastal strip.) In total, 86 million cubic metres of material was excavated and carried by conveyor belt to the coast for loading onto special bottom-opening barges that made the 20 km. journey to unload the filling material at the site of the artificial island. At the site the material was worked and levelled to a height of 5 m. above sea level; this gave the island an average thickness of 12 m. The artificial structure rested on a natural sea-bed of alluvial clay overlaying sand and pebbles and the whole was consolidated by driving piles through to the sand and pebble layer. The wharves and breakwaters which constituted the margin of the island were made of concrete *caissons* and would withstand waves up to 4.5 m. high.

The Kobe Port Island scheme construction was completed in the early 1980s. It was an ambitious project in that it sought to include features other than solely industrial ones and as such it represents one of the earliest attempts to create a complete range of community activities on new land off-shore. Whether the residential, recreational and commercial aspects prove as successful as the industrial one will become evident some time in the future.

EXAMPLES OF ARTIFICIAL ISLAND PROJECTS—3: OIL EXPLORATION IN THE ARCTIC

Much of the earliest development of fixed and floating platforms has been undertaken in response to the oil industry's need to pursue oil exploration at sea. Now, because of the unique conditions of oil exploration in the Arctic, the oil companies are also interested in a variety of artificial islands for shallow water sites which can withstand the forces exerted by the movements of pack ice. In some of the sheltered sites in the south of the exploration region there is sufficient time free of ice for conventional drilling ships to be used and for more or less standard rigs to be used for production. In some cases these will require additional strengthening and stabilizing to cope with ice, tides and currents. Several designs achieve this. One type is a fixed island similar to a conventional rig but with massive *caisson* feet. Drilling was carried out down through the middle of one of the legs in order to protect the drill apparatus. *Caisson*-strengthened rigs have been used in up to 40 m. of water. Another option for harsher conditions is the 'mono-pod' platform. More suited to shallow sites, it consists of a single column, some 10 m. in diameter, which supports the platform. Once again, all the operations are carried out down the column where there is protection from ice.

Variations on the conventional fixed platform have served for Arctic oil exploration very successfully. But as operations moved yet farther north it became impractical to rely on conventional designs however they might be reinforced. Attention then turned to artificial gravel islands for shallow water sites on the northern edges of exploration. These islands are created, quite simply, by dumping gravel at the site. In summer the gravel can be dredged from the seabed, but in winter it has to be excavated from land sites and transported. The gravel island rests directly on the sea-bed and has a very wide base; together with the large mass of gravel, this makes the island extremely resistant to the force of the surrounding ice. It thus provides a stable platform for year-round operations. Initially the islands can be constructed on a scale suitable for exploratory drilling, but if necessary they can be extended to carry permanent production in the event of oil being discovered.

In respect of the costs usually associated with oil exploration, gravel islands are cheap to construct, especially if the gravel can be found in the vicinity of the intended island site. They can be built to the absolute minimum specification necessary for short-term exploratory drilling to be completed. If a strike is made the island can be consolidated and extended for production, but if no oil is found the island can simply be abandoned. Once maintenance is withdrawn, the island will gradually be eroded away, so there are no costs incurred from dismantling a site after drilling has been completed.[6]

The shallow sloping sides of gravel islands (which result from the natural settlement of the dumped gravel) mean that each small increase in the depth of water requires a far more than proportionate increase in filling material. Therefore, as gravel islands edge into deeper water it will probably be necessary to employ a steel shell or *caisson* ring to limit the diameter of the island at the sea-bed; otherwise the quantity of gravel needed, particularly if it has to be transported to the site, will be prohibitive. Certainly for larger and permanent production islands a *caisson*-retained fill construction has considerable advantages both in terms of reduced maintenance costs over the longer period of operational life and of initial construction costs if fill material is not locally available.

Since the mid-1970s several companies, including some of the largest, have been involved in building or operating gravel islands in the Arctic seas. Construction techniques vary in detail, but the general principle is to lay a foundation layer of gravel and sand. This can be either be dumped directly out of a barge, which gives a very compact mass but with very shallowly sloping sides, or it can be deposited through a pipe hanging vertically from the barge to just above the sea-bed. This latter technique provides more control in laying the material and, because of the lower velocity of the particles, allowed a steeper slope to form with consequent savings in fill materials.[17] When the sand/gravel base has been laid to between 7 and 10 m. below the surface a top layer of gravel is added. The mound is levelled and consolidated before the working platform is put in place. The deck may be of steel or concrete and in some designs ships' hulls have been converted to serve as the platform. Normally the top of the deck is between 10 and 15 m. above sea level, all the drilling facilities are carried on this deck, which can be removed and used elsewhere if the island is abandoned. One of the largest islands of this construction is Esso's Issugnak installation which is 26 km. off-shore in water 19 m. deep.[8] In 1980 it was extended and the working area now has a diameter of 135 m.

For the future, and even deeper water locations, *caisson* ring fill islands will be built. Structurally, the important feature is to ensure the integrity of the *caisson* ring against the ice. In a sense, the return to *caisson* shell negates the advantages of gravel islands for withstanding the pressure of ice movements, but in deep water cost will militate against simple gravel islands. A combination island of gravel fill with *caisson* shell embankments should provide the compromise between stability and cost for deep water sites.

ARTIFICIAL ISLAND PROJECTS IN THE FUTURE

The examples above are all of tried and tested programmes, some of which have been used for more than a decade. In the future the constructional practices of artificial island development may not alter radically, but the range of uses to which they will be put and the scale of those uses may increase dramatically.

The Japanese already have Nagasaki international airport, opened in 1975, operating on reclaimed land off-shore and many countries have airports where runway extensions have had to be built out to sea because no land was available elsewhere.[9] One ambitious offshore airport project is that for Kansai airport in Osaka Bay which would serve Japan's second city. The environmental and political issues of new airport development in Japan have led to serious social protest. Siting airports off-shore may forestall further protest, remove noise pollution and, of course, conserve land on shore. The Kansai site is 5 km. off-shore in an average of 18 m. of water. If given clearance, construction will begin in the 1990s (using techniques similar to those employed at Kobe Port Island) of three 4 km runways, the first of which could be open very soon once construction has begun.[10]

Rather more distant are proposals for fully-floating airports. A floating STOL (short take-off and landing) port has been designed for the North Sea to receive STOL passenger aircraft. As proposed, it would have three decks. The top one is the runway, which measures 600 m. by 70 m. The middle deck is for passenger handling and the lowest deck contains hangar space for maintenance facilities. The greatest advantage of the floating airport is that it can always be aligned into the wind for take-offs and landings. The most important feature of the design must be the minimization of the effects of wave action. The conventional answer to problems of stability is a semi-submersible structure with large floating pontoons below sea level. The platform is thus isolated from the waves and other surface disturbances. This solution would be a suitable one, but a more radical option is the 'air cushion'. The platform becomes the top surface of a cylinder, the open-ended bottom of which is below the water. The design is sometimes described as a floating gas holder. A cushion of air is trapped between the platform and the sea and this cushion absorbs wave energy to provide the platform with the additional stability that a STOL port ideally requires. Clearly, the case of the North Sea implies that the impetus for development is from the oil industry, but small, mobile STOL installations would find opportunities for deployment in many other locations.

Full-sized floating international airports have also been proposed. A speculative draft design has been drawn up for two parallel 4,250 m. runways with taxiways and aprons. The platform could be made of hollow concrete slabs filled with expanded foam plastic. Construction costs are greater than for building, from scratch, on land, but the engineering is not beyond current practices and, given the current climate of opinion in, say, Japan, a floating international airport is not an impossibility early in the next century.

Projects such as Kobe Port Island have demonstrated the practicality of extending industrial facilities onto artificial islands. Extensive proposals for the North Sea were produced during the 1970s by a consortium called the North Sea Islands Group. Sites were suggested several kilometres off-shore from both East Anglia and the Hook of Holland in water between 20 and 30 m. in depth. Construction would be of *caisson* and sand filling and would take as much as fifteen years to complete. Not only this, but the severe weather in the North Sea inevitably means high maintenance costs throughout the structure's operational life. The two other major drawbacks to the scheme were firstly the expense and the logistical difficulties of transporting the whole labour-force to and from the site daily (since these sites were not conceived as offering accommodation) and secondly the simple problem of siting the island in one of the world's busiest seaways. In the event, subsequent economic conditions in the West have given no impetus to the scheme and no progress is likely in the next few years. However, the existing experience in North Sea operations and the possibility that Western nations may become increasingly keen to locate chemical and heavy engineering industries away from areas of habitation leaves the North Sea as one of the most likely sites for extended artificial island development.

One obvious candidate for an artificial island is the nuclear power industry! In the United States a programme was begun in the early 1970s with the formation of Offshore Power Systems Inc., which was a consortium of corporations already involved in power generation. The consortium proposed a twin reactor system to be sited 4.5 km. off New Jersey. In 1972 the reactors were actually ordered, but there were delays in licensing and the subsequent reduction in the rate of growth of electricity demand resulted in cancellation by 1978. However, in 1982 United States regulatory bodies granted outline approval for offshore nuclear facilities and at much the same time Westinghouse (one of the members of the original consortium) claimed that a 1150 MWe generating plant built off-shore would be cheaper than one built on land. However, no progress is to be expected unless the demand for electricity shows increases greater than those of the last decade. The European experience has been similar to that in America. Studies begun in the early 1970s found the proposal superficially attractive, but by the time reports were completed several years later conditions had changed and offshore sites were deemed unnecessary as either a short or a medium-term option. While concurring with American and European conclusions, the Japanese have also been investigating offshore fossil fuel power stations, partly as an end in their own right, but also as a step towards offshore nuclear reactors, should that option become attractive again in the future.

Many of the marine technologies discussed elsewhere in this book either presuppose the development of artificial island technology or would, at least, be considerably advanced by such developments. An OTEC plant is, in effect, a floating artificial island, and the mining of manganese and phosphorite nodules could be made economically more attractive by the provision of processing facilities at the mine-site.

One final example will illustrate the range of structures that could serve as artificial islands and the extent of opportunities available for the use of materials other than steel and concrete. It has not yet been necessary to look beyond conventional materials and construction techniques, but one alternative material with interesting potential is ice. Once again, this idea owes its development to the offshore oil industry as an option for short-term exploratory drilling platforms in the Arctic. The principle is to spray water over an ice sheet in order to increase its thickness. The spraying begins early in the Arctic winter and continues until the sheet is of the intended thickness. In 1977 Union Oil carried out drilling between January and April from an ice island grounded in 2.7 m. of water. The surface of the island was raised one metre above sea level and this gave sufficient protection for the operations. Once drilling was completed the equipment was removed and the island had melted by July. Ice islands are lighter than gravel ones and less resistant to the lateral movement of surrounding ice. To overcome this, the Union Oil island was protected by a moat cut into the ice around it. A subsequent ice island of the late 1970s, again for shallow water, had a surface raised 6 m. above sea level and that size of island was found to be able to resist the forces of the surrounding ice even during the spring thaw. However, despite its size, the island deteriorated in open water during summer and it was evident that ice islands would not survive to a second winter's operation without insulation and extensive maintenance. As yet there is no incentive for ice islands to be used for other than very short-term work and so long-term durability is not justified in terms of the expense of maintenance. Ice islands are one of the few genuinely novel ideas for artificial island construction. Their use is obviously restricted and even in the Arctic they are as much a novelty as anything else. But ice islands do demonstrate that steel, concrete and conventional construction practices do not exhaust the possibilities for artificial islands even if they have dominated up to now.

COMMENTARY

We have seen the wide range of structures and uses for artificial islands, ranging from the fixed platforms of the offshore oil industry through to proposals for floating international airports and a variety of specific proposals in between. All those projects currently in operation and most of those proposed for the future (the ice islands being the obvious exception) are conceived within the scope of conventional marine construction. Even the largest-scale proposals only require the extension of existing practices. Of all the marine technologies discussed in this book it is artificial islands that demand the least in terms of technical development for their projects to proceed.

The use of artificial islands will, no doubt, piecemeal increase steadily in the coming decades in response to the needs of specific situations. In general it seems that in Europe and North America the main impetus for development will be the desire to remove to sites remote from human habitation those industrial operations which are noisy or dirty or with a high risk factor. Increasing legislation may make offshore sites politically attractive even if costs may be initially higher. Scarcity of land for industrial expansion is unlikely to be a major factor in the West; however, in the Far East the shortage of space, particularly in coastal locations, will fuel the drive off-shore. Since the Japanese need to expand industrial capacity is greater at present than the West's desire to relocate its heavy industry, it is the Japanese who have been most active across a whole range of artificial island schemes. Where neither space nor relocation are factors there is a third circumstance in which artificial islands will begin to be employed more extensively. At sites which are inaccessible or at which conventional building of plant and services would be prohibitively expensive, complete factories or process plants can be provided on artificial islands. Such options apply not only to the supply of plant to developing nations, where we have seen that the provision of complete prefabricated installations may create political and social problems, but also to the exploitation of the mineral and energy resources of the deep oceans, where the provision of adequate floating installations is central to the success of many deep ocean projects.

For all categories of artificial islands, however, the legal framework requires clarification before widespread developments can be expected. The legal status of artificial islands is confusing with such a variety of designs, locations and uses. In some instances an artificial island can be treated as a ship or as a natural island and here precedents do exist, but this is increasingly unsatisfactory and a new regime is needed. It must take account of the variety of artificial islands, in particular of those hybrids which cannot be classified simply as a ship or as an island, and it must recognize that the rights and responsibilities of the builders and operators of artificial islands may be different in a country's offshore economic zone than on the high seas. Furthermore, legislation must deal with questions of safety, especially since artificial islands may be sited in busy seaways. One safety aspect that is certain to need legislation is the prevention of environmental damage and the apportioning of responsibility if such damage does occur. Just because high-risk and polluting activities are located at sea, this does not mean that environmental damage at sea is sanctioned. Rather, it means that if there are accidents at least their immediate impact will be away from land. Legislation will need to ensure that adequate environmental safeguards are in operation even at isolated locations. Some of these issues are implicit in current negotiations on the law of the sea under UNCLOS, others may not be resolved in that forum, but any clarification in the legal status of artificial islands should provide a clear impetus to their increased construction and operation.

NOTES

1. They are not suitable, of course, for operation in shallow water.
2. Some fixed oil platforms are now being deployed in deeper water, but fixed islands will always be restricted to the continental shelf.
3. In *polder* reclamation the area to be reclaimed would normally be adjoining dry land, but this is not necessarily the case for fill islands which could be entirely separate from the shore.
4. One of the first large-scale reclamation projects in Japan was the international airport at Nagasaki, opened in 1975. This was built only a little way off-shore; other projects are planned to go much farther away from the shore.
5. By 1979 the harbour reached the projected figure of 18 million tonnes of freight per year. Some of the berths were specially designed for oil and other high-risk cargoes.
6. This ability for self-demolition is also environmentally quite useful. Not only is the gravel itself unproblematic, but the gradual decay of the island is preferable to the sudden removal of a more permanent steel or concrete structure.
7. The steepest slope usually obtained has a gradient of 1 in 5.
8. This is not particularly deep. Many operate in much deeper water.
9. For example, Hong Kong, La Guardia (New York), Nice and Kingston (Jamaica).
10. A very similar plan for a new international airport for Hong Kong awaits the implications of recent Anglo–Chinese agreements on the colony's future.

8 Surveying and sensing for marine technology

Implicit in the development of new marine resource technologies is the need for parallel advances in a whole range of related practices. The creation of this technical infrastructure is essential for the realization of the new technologies. Many elements of the infrastructure are neither unique nor exclusively relevant to resource technologies. Indeed, on the contrary, developments in navigation, construction and maintenance, safety at sea and other general maritime operations are all more immediately applicable to existing shipping, fishing and offshore hydrocarbon industries. Of subsequent importance though these advances may be to new resource technologies, the impetus for initial development is largely independent of them. However, one set of developments is more particularly related to new technology and this is the field of sensing and surveying in the marine environment. Simply stated, this is concerned with locating, mapping and assessing the extent of marine resources. These techniques are principally applied to the surveying of mineral resources and hence are analogous to conventional mineral exploration on land, but some sensing techniques are available which provide oceanographic information on tides, currents, ocean temperatures, etc. which are relevant to the assessment of ocean energy resources.

The importance of accurate surveying is clear. Only when resources (mineral and energy) have been located and evaluated is it possible to estimate the value of those resources and to begin to delineate the economics of exploiting a given resource. From some of the case studies it is clear that a major cause of uncertainty in assessing the commercial prospects for marine resources is the absence of reliable survey data. Any progress which improves the information available will assist in clarifying the decision-making process for new marine technology. The purpose of surveying and sensing techniques is not solely the economic evaluation of the resources. They also permit the accurate location, and subsequent relocation, of resources. Clearly, any surveying depends on accurate positioning and mapping, so that surveying and sensing techniques are closely associated with the related techniques of navigation and station-keeping. Accordingly, some surveying and sensing techniques are principally concerned with location, while others are more suited to the mapping of sea-bed topography. This is not to say that the techniques are mutually exclusive: there is a certain degree of overlap between them. Existing methods of marine surveying and sensing are diverse and are described briefly below.

Of those techniques primarily intended for navigation and positioning there are three major categories. Firstly, radio position fixing systems permit a location to be determined with reference to the position of known radio transmitters. There are a variety of radio fixing systems available, depending on the specific needs of operations. Low-frequency systems can be used over long distances but they do not give the accuracy of high-frequency systems. Unfortunately, the higher the frequency, the shorter the range of operation. Indeed, with very high-frequency microwave systems operation is virtually limited to the line of sight. The second position fixing method is a sea-bed acoustic position fixing. A number of acoustic transponders are placed at accurately known locations on the sea-bed and then new positions can be precisely determined relative to those lcoations. The range of transponders is of the order of a few kilometres and they enable very accurate fixings on the sea-bed. The third system is now becoming increasingly important for all navigation and location fixing requirements at sea: it is the use of satellites. Satellite navigation systems permit a precise, global network to be established for all maritime operations. All three of the above systems provide accurate fixes for mapping and surveying marine resources; they are not, however, principally survey techniques in themselves.

There are many survey techniques available, depending on the task to be undertaken. They can be divided into two broad categories; acoustic methods and field methods. The former are still extremely important for surveying at sea. They all operate by analysing the reflection of a signal from the structure being surveyed. The latter methods do not require a signal to be generated for reflection; instead they measure small variations in the earth's gravitational and magnetic fields. These can then be related to specific geological formations.

The simplest acoustic method is echo-sounding. This gives a measure of water depth by the time of the delay between the generated pulse and the return echo, and hence sea-bed topography can be mapped. Side sonar scanning is similar in principle to echo-sounding except that the echo-sounders are tilted away from the vertical. This slanted sonar produces a complex record of the reflection, which can be subject to considerable distortion unless the equipment is towed relatively close to the sea-bed. Accordingly, interpretation of side-scan sonar records is not always straightforward. However, in conjunction with sampling, it has proved a useful tool in mapping the distribution of sediment on the sea-bed. A third variation of echo-sounding permits reflections to be received from beneath the sea-bed. At low frequencies the pulse will penetrate the sea-bed and produce reflections from the underlying geological features. Depending on the frequency and the energy of the pulse, reflections can be recorded from varying depths beneath the sea-bed.

All three echo-sounding techniques require the same basic apparatus: an acoustic source, an acoustic receiver and a recording system. The precise choice of equipment depends on the specific application intended, but the choice of acoustic source is the main variable. The source should produce a pulse of high energy and very short duration. High-frequency pulses render improved resolution, while low-frequency pulses permit longer-range operations or penetration of the sea-bed, but at the expense of resolution.

The primary marine uses of techniques derived from echo-sounding are general mapping of sea-bed features, mapping of surface deposits and broad mapping of features below, though relatively near, the sea-bed.[1]

For deeper exploration, another, not dissimilar, technique is used: seismic reflection. This was the method extensively employed in early exploration of the North Sea hydrocarbon deposits. It requires the generation of a very high-energy seismic shock wave whose reflection is then recorded as in the other acoustic methods. The very high energy is produced in a number of ways, including the implosion of an artificially created vacuum cavity in the water and even the use of chemical explosives.[2] This enables surveying to be carried out at depths of up to 5 km. Clearly, the resolution is not as good as for the shorter-range echo-sounding technique, but the results are very important for identifying large-scale features such as those associated with oil and gas reservoirs. The record of reflections from seismic reflection surveys are very complex. Multi-channel recordings are taken from up to 200 hydrophones and several 'shots' are taken for a given depth. The results need to be analysed and amalgamated with the aid of computers before any pattern can begin to emerge.

The second category of techniques was the field methods. Unlike the acoustic methods these do not require the generation of a signal for reflection; instead they measure the existing magnetic and gravitational fields. Small variations in these fields can be interpreted in terms of the geology of the area surveyed. Magnetic surveying can yield very accurate information and is suitable for mapping either small areas in detail or very large ones more generally. The instruments are usually carried on board ship or in aircraft and extensive marine surveying has already been undertaken using this method. Gravimetric surveys at sea have much more limited accuracy than those on land. Consequently, the method is only suitable for surveying major geological features and structural discontinuities. Both field methods are normally used in conjunction with acoustic methods.

Acoustic and field surveying techniques are primarily methods of surveying from within the marine environment. The equipment is usually carried by a vessel on or in the sea. Future developments will increasingly see the use of remote sensing for marine surveys. This means sensing of the marine environment from outside it: in practice this means above it. This can be achieved from aircraft, as indeed some magnetic field surveys already employ, but more likely is the development of surveys from space. There are several advantages to be gained from satellites as opposed to aircraft for remote sensing.

The combination of the photographic camera and the aeroplane has for many years provided high-resolution information about the earth's surface. However, this arrangement is severely limited if more detailed and varied information is required. First, with an increase in the desired area of coverage so the aircraft operational costs will increase, but without the use of many aircraft, the point of sensing will vary substantially across the area. This lack of uniformity may in turn limit the reliability of the complete picture, as in the case of analysing weather or wave current movements. Secondly, it may be desirable to measure parameters other than those seen by visible light, such as temperature detection by infra-red radiation.

The advantage of satellite sensing comes from the potential for covering large areas of sea or land in relatively short periods, and hence producing a wide-ranging coherent picture, at reasonable cost. The effectiveness of the aeroplane versus the two types of satellite is set out in Table 8.1.

Remote sensing by satellite, of the land, is at a much more advanced stage than that for the oceans but it has acted as a precursor to the technological and interpretive development of resource remote sensing of the oceans. Satellites may be used as discrete instruments offering cost and time advantages compared to traditional data gathering methods. They may prove more accurate and may perform hitherto impossible functions.

Remote sensing by satellite of the marine environment usually refers to the acquisition and analysis of electromagnetic radiation in the visible, infra-red or microwave parts of the spectrum, emanating either naturally or by reflection from the surface of the earth. The sensors themselves are

Table 8.1 Aeroplane effectiveness as against two types of satellite

Platform	Height (km)	Circular area under view	
		Km^2	% of the earth
High-flying aeroplane	15	235	5×10^{-5}
Orbiting satellite	1,000	7.3×10^5	0.15
Geostationary satellite	36,000	2.5×10^7	5.0

described as *passive* if the energy they detect is radiated naturally from the surface being observed and *active* if the instrument itself irradiates the surface, for instance using lasers or radar and detecting the backscattered energy. It is the use of surveying by electromagnetic radiation that permits satellite remote sensing to give information about currents, wave patterns, sea temperatures, etc. and hence to allow surveying of resources other than the conventional physical resources that are the targets of acoustic and field survey methods.

Compared to satellite sensing over land, the marine environment presents considerable problems of interpretation and ocean surveying is still only in the early stages of development. Nevertheless, the potential of the techniques, once perfected, encourages further efforts, and to this end several programmes are currently running.

The first two decisions to be taken concerning a remote sensing satellite are the choices of orbit and instrumentation. To follow a consistent orbit, satellites must fly several hundred kilometres above the ground, usually 500 to 1,000 km. but at the same time this figure must be reduced as far as possible to maximize the resolving power of the sensors. The orbital path is selected according to the tasks to be performed, but in essence this involves a choice between two alternatives. The path may be over or close to the poles or around the equator. Best spatial coverage is achieved from a polar orbit, since in the time required to complete one orbit the earth will have rotated a few degrees to the east, thereby ensuring that the next orbit will travel over a different path of the earth. There is then, however, the problem of securing total coverage. With the separation of the orbits at the equator likely to be several thousand kilometres and the footprint of the sensor (that is the width of the swathe scanned at any instant) typically only hundreds of kilometres, large tracts between successive orbits will remain unsurveyed. To avoid this problem, the speed of the satellite is adjusted so that the number of orbits completed in a twenty-four-hour period is not an exact number. By doing this the satellite does not arrive back to cover the point at which it started after twenty-four hours, but is offset by a few hundred kilometres. Provided that the swathe width is greater than the offset, the satellite orbit path will ensure that the sensors cover the whole earth in a number of days. Further adjustments of the orbit with respect to the equator and the time to complete an orbit render it possible to synchronize the orbit time with the earth's rotation, ensuring that the satellite passes over every point of the earth in the daylight part of its orbit at the same time of day. Similarly, every point of the earth can be covered in the night-time part of the orbit and at a common time. This type of orbit ensures uniformity of data, which simplifies the analytical task.

As an alternative to the polar orbit, the satellite can be positioned in an equatorial orbit travelling at such a speed that it takes exactly twenty-four hours to complete one orbit. The satellite circles the earth at the rate at which the earth spins on its axis and it therefore appears to be stationary and can survey one part of the globe without interruption. Satellites in geostationary orbits have to be placed at a height of approximately 36,000 km. above the equator and as a result the sensor resolution is poorer than with the lower flying polar orbiters. In addition, geostationary satellites suffer the disadvantage of day and night cycles and cannot adequately cover the polar regions. However, they do have the major advantage of covering large areas of sea and the ability to continuously scan the same point on earth. Programmes such as METEOSAT use this concept to provide a world-wide synoptic view of the weather.

The choice of instrumentation again depends on the intended use. As previously stated, satellite remote sensings study radiation emitted or reflected. According to the source of radiation being measured, the techniques involved can be divided into three categories.

(a) passive techniques where the sensors detect arriving radiation that is emitted by the body itself,
(b) passive techniques which make use of sunlight reflected from the body under observation,
(c) active techniques involving sensors which emit an artificial source of radiation such as radar waves, then receive and analyse the reflected signal.

Types of sensors are broadly divided into non-imaging and imaging groups. The first of these are not able to scan within small areas over the field of view. For example, they may detect the absence or presence of a gas or, alternatively, they may quantify a parameter as a whole over the viewable area. These limitations restrict the usefulness of these sensors in the context of oceanographic remote sensing.

The use of imaging sensors provide the ability to scan the field of view and measure the different intensities of radiation within frequency bands. It is therefore possible to build up 'pictures' of the earth for particular frequency bands. Consequently, the imaging sensor has proved most useful with environmental remote sensing.

One further important dimension, defining the operation of imaging sensors, is resolution at the ground. This is determined by the design of the sensor and by the speed and flying height of the satellite. For typical satellites, ground resolutions will vary between 25 m. and several kilometres—the expected use determining the particular performance in each case. For example, when assessing the wind fields over the oceans a resolution of 1 or 2 km. is adequate.

The major sensing techniques are the various forms of optical scanners which are passive instruments and the active microwave sensors. The former are scanning instruments which operate in the optical or infra-red wavebands. The exact frequency for sensing is determined by the use of filters. Results at several different frequencies can be combined to produce a composite picture. These detectors are important for determining water temperature and for showing small colour changes in shallow waters, which is significant in locating discharges, such as those from damaged pipelines into coastal waters. The latter group of sensors are the active type. Using microwave sources, they are analogous to radar and provide data concerning the surface of the sea. Very exact measurements of the distance between satellite and surface can be produced; the wave conditions can thus be inferred and[3] by analysing the scatter of the returning signal, information can be gained about wind and current patterns at the surface.

Given existing technology, there are a range of ocean parameters that can be directly measured by remote sensing, or where measurement can be derived. The levels of accuracy will vary substantially according to sensor technology and data interpretation. In some areas the methods of detection and measurement could be described as mainly experimental. In addition, the movement towards microwave sensing will depend on developing new methods for data analysis and interpretation, even for those parameters where effective passive techniques exist. Established areas of expertise include measuring sea temperature, pollution monitoring, and the detection of ice and icebergs, while the observation of features of ocean waves and current is a growing area for microwave sensing.

It is clear that remote sensing is not principally a method of conventional surveying for mineral resources. In so far as it is directly relevant to the assessment of marine resources, it provides data about the energy resources of ocean temperatures, waves and currents. But it also relays important information on the ocean environment, which will be crucial in determining the sites of deep ocean installations and the conditions in which they must operate. Perhaps, therefore, it is unsurprising that the origins of remote sensing are found in the operation of satellites designed for meteorological tasks, in particular the weather satellite programme of the United States National Oceanographic and Atmospheric Administration: but the first dedicated oceanographic satellite was NASA's 'Seasat A', launched in mid-1978 into a near polar orbit at 800 km. The satellite had an orbital period of approximately one hundred minutes and completely scanned the globe about once every thirty-six hours. Planned as a proof of concept mission, aimed at determining the most important features of an operational system, the satellite operated successfully for 106 days, completing 1,503 revs, before the failure of the prime power system on 9 October 1978. On 21 November 1978, NASA officially declared the satellite lost. The experiment cost about $100 million.

'Seasat A' carried five sensors (three of which were active) and these provided data on wave heights, ocean tides and surges, currents, wind direction and speed, ocean temperature and distribution of ice flows. Despite its brief period of operation and its mechanical failure, 'Seasat A' was successful in demonstrating the concept of oceanographic survey from satellites. Some apparatus performed especially well and wind speeds were determined to within 12 metres per second and sea surface temperature to within 1.5 °C.

Subsequently, oceanographic survey has been included in various other satellite programmes, whose primary objectives have usually been meteorological. The nearest to a specifically marine resources satellite programme has been, despite its name, the 'Landsat' series. This was an earth resources programme begun in the 1970s which has provided considerable data on both coastal and ocean waters. Technical problems in the early 1980s caused delays in the 'Landsat' project. Indeed, the future of several American satellite projects is at present unclear since NASA's recent loss of several satellites launched both conventionally and from a space shuttle. The development of oceanographic remote sensing may be adversely affected in consequence of NASA's problems.

In the event that American projects are either delayed or even curtailed, other countries are beginning to consider the inclusion of marine sensing in their own satellite programmes. A joint European and Canadian consortium has planned the ERS-1 project, which should be launched towards the end of the decade. Although designed for a range of geographical and meteorologial surveying tasks, ERS-1 will undertake extensive remote sensing of coastal, ocean and polar regions. It will be primarily concerned with wind speed measurements and recording wave and current patterns.

One Japanese programme scheduled for the middle and later years of the decade is the MOS (Marine Observation Satellite) project. It is intended to include three satellites which again, will concentrate on wind and wave observations, but the satellite will also measure changes in sea surface temperature and changes in the colour of the ocean. The second and third satellites are planned to carry more sophisticated sensing equipment.

COMMENTARY

Remote sensing of the oceans from satellites is still in the early stages of development. Its progress so far has been largely as a consequence of programmes whose aims were primarily meteorological. However, during the 1980s oceanographic resource studies as such began to be included as major goals of satellite projects.

Satellite sensing has two important benefits for new marine resource technologies. Firstly, it provides the possibility of deriving estimates of the energy resources of the oceans, since it offers comprehensive and extensive data on ocean temperature, wave structure and wind and current patterns. From these, much better estimates of the potential energy resources of the oceans should be available. Secondly, remote sensing is providing much data on the marine environment, especially in the open ocean, and this information is vital to an understanding of the conditions in which deep ocean programmes will need to operate.

The future for remote sensing is probably more assured than almost any of the specific marine technologies themselves since it will continue to be developed in association with meteorological observations of the oceans. Marine resource surveys might not be given a high priority in the goals of weather satellite missions, but so long as such satellites continue to be launched there will be a continuous supply of data on which interpretations of marine resource availability can be made and analytical techniques can be improved. Obviously there is some question over the short-term future of American satellite programmes in view of NASA's problems with satellite launching in the mid-1980s. This is unlikely to be a long-term threat to remote sensing.

It is in the area of remote sensing, rather than in any of the other new marine technologies, that an overt military aspect can most obviously be implied. There is no doubt that all satellite surveillance will attract military interest and this may ensure that finance will remain available for the support of remote sensing projects even if those projects are not directly under military control. Whatever aspect of surveillance is being developed, its techniques and apparatus may eventually find an application in the surveying of marine resources.

The successful development of satellite surveying and sensing techniques for the marine environment prove to be the crucial technical advance that will enable a whole range of other marine technologies to proceed. Remote sensing offers the prospect of accurate resource estimates which are crucial if commercial decisions are to be taken on the future of new marine technologies.

NOTES

1. For marine technology the mapping of the sea-floor is important not only as an indication of sea-bed mineral resources but also for the information which is needed for siting platforms or artificial islands on the sea-floor.
2. The energy of the signal is of the order of several tens of kilogrammes of TNT.
3. The height of waves can be estimated to within one metre.

9 Summary and prospects

SUMMARY

After considering the range of new technologies that have been proposed for the extraction of mineral and energy resources from the oceans it is not unreasonable to ask if there is any justification for claiming the existence of a generic 'marine technology'. Diversity rather than unity appears to be the major characteristic of the projects and it is an impression we have sought to emphasize by the importance we have attached to the specific features of each individual case study. Certainly the technologies have the marine environment in common. They would all be sited in, on, under or at the margins of the oceans and they must all face the problems of operating within that environment. Severe engineering challenges would be posed by the construction, operation and maintenance of marine technologies, especially, but not exclusively, by those sited in the open waters of the deep mid-oceans. But these challenges specifically associated with the maritime environment are essentially technical ones and throughout the case studies it has been the intention to demonstrate that technical questions are only a part of the puzzle for which marine technologies must find answers.

So, beyond the technical necessity to master the marine environment, do different schemes possess much in common? At first sight it is not clear that they do. The arrays which would harness the energy of ocean currents are conceived on a huge scale, covering many square kilometres of ocean, whereas Red Sea metalliferous muds can be mined on a small scale by a single-surface vessel. The proposal to extract energy from salinity or density gradients in the ocean sounds almost unbelievably exotic and is in total contrast to the prosaic dredging of phosphorite nodules for direct use on the land as fertilizer. Technically, the design and operation of devices to generate electricity from the motion of waves is simple and has been capable of demonstration for many years, but tapping directly onto the supersaturated metal-rich brines from hydrothermal vents on the floor of the deep oceans is technically way beyond current practice. Artificial islands may be appropriate for many locations throughout the world, while tidal barrages are restricted to a handful of sites. Minerals dissolved in water can already be extracted on a commercially viable basis from the mineral-rich waters of the Dead Sea by evaporative methods, but similar commercial extraction from seawater awaits the development of ion exchange and a more favourable world market for the metals it would produce. World markets and simple commercial economics had little to do with the early support of and impetus for American ocean energy programmes such as OTEC and Dam Atol. These programmes were supported for strategic reasons involving the politics of energy resources. In the event OTEC carried the day politically, so leaving the Dam Atol and other schemes to founder. The expansion of dredging for gravel and sands on the continental shelf will raise questions concerning the environmental impact of large-scale dredging in shallow waters, whereas the recovery of manganese nodules and radiolarian oozes from the deep oceans would have considerable importance in the context of international negotiations on the Law of the Sea; in the deep oceans the environmental consequences of dredging do not seem to generate the same concerns as when inshore sites are involved. It would be possible to continue contrasting these schemes one with another, dividing them up again and again according to their differing characteristics and emphasizing the diversity within the term 'marine technology'. But the point need not be laboured further. It should be clear that each project contains its own, perhaps unique, constellation of technical, commercial, geographical, political and environmental features. This alone warns against seeking for an 'identikit' model of new marine technologies or for any simple algorithm by which to assess the future prospects of marine technology.

A second look at the case studies however, gives some cause for renewed hope for making sense of the pattern of development in marine technologies. The pattern is certainly not simple, but neither is it random. For although each project might appear to be a singular combination of factors, these factors are more or less common to all the programmes. The combination of these factors and their relative importance in each case varies, but the same fundamental issues are common to many of the marine technology proposals. One has only to return to the case studies to see that many of the themes recur in otherwise superficially unrelated projects, but each theme is varied by the specific constraints and context of the individual situation. More detailed study of the cases shows there to be seven distinct central themes which provide the core that characterizes the recent history of marine technology developments. Each of these seven themes are discussed (in no particular order) below.

In the first place it has been quite clear that technical competence, while necessary, is by no means a sufficient criterion for a new technology to proceed. This is perhaps obvious, yet much of what was written in the 1970s concentrated almost exclusively on the technical requirements of the new technologies. Little concern was shown for the possibility that other factors might be important, especially in regard to the commercial prospects of a programme. This is not surprising since these schemes were still in their R & D phases in the 1970s. By definition, R & D is almost wholly concerned with technical matters and the solution of technical problems is enough to sustain a research programme. But once the project begins to move out of R & D alone, into the phase which might lead to commercialization, technical virtuosity is an insufficient criterion on which to assess its future. During the 1980s many of the marine technology projects reached the stage where they had to leave the laboratory and produce pilot plants prior to commercialization. At that point it became a serious error to continue to perceive the progress of the projects in primarily technical terms since the technical parameters were no longer the whole story. The R & D programmes were apparently very successful in most cases and rarely were unsolved technical problems responsible for failure to achieve commercialization. Indeed, the overall technical success of the marine technology programme has been most impressive.

The second element is that the level of intrinsic commercial viability is important if a scheme is to be developed beyond pilot plant stage. Strategic considerations seem able to support R & D programmes, which are relatively cheap compared to the cost of full-sale deployment, but when it comes to the question of whether or not to proceed with commercial development a clear case for the viability of the project becomes much more important before the commitment can be made. The case for the commercial viability has rarely been forthcoming; despite the accompanying rhetoric the case for the strategic development of ocean energy and mineral resources has not been borne out in practice. Even in Japan strategic arguments for the development of new sources of minerals and energy are only applied as far as the completion of an R & D programme. Beyond that, most projects are expected to justify themselves on commercial grounds if they are to continue after initial government finance has been withdrawn. Japan, which of all countries probably feels the greatest need to secure for itself supplies of energy and raw materials that are less vulnerable to price rises and interruptions, has not yet shown itself prepared to support new schemes regardless of cost. Indeed, worldwide, that has been the rule. New marine resource projects have been in the main unable to offer convincing evidence of commercial viability and it seems that although strategic considerations might permit a market for resources of marine origin which were slightly more expensive than conventional supplies, few programmes will be able to meet this criterion given existing market prices. There is no evidence that any country rates the strategic value of marine-originated resources so highly as to be prepared to pay a premium way above current market prices. Many of the marine schemes came to the fore in the late 1960s and early 1970s when demand and prices were rising for a range of raw materials. Political uncertainty over the security of supplies and extrapolations which predicted resource shortages by the end of the century all suggested that alternative sources would not have to be all that cheap to find a niche in world supply. In contrast to this scenario, the last decade has seen many resource prices depressed and the addition of alternative supplies would only depress prices further. The commercial aspects of the new technologies have therefore been restricted less by their own failures to produce at prices competitive with conventional supplies than by the failure of general resource prices to rise to the levels which were anticipated. Nevertheless, the consequence is that marine resources are unable to compete and strategic support for them is insufficient to redress the balance.

The third theme is that of political commitment to the projects. In Japan and in the United States, especially under President Carter, marine technology development has received government support. In the United States this political support was achieved through federal financing of research projects, while in Japan, apart from direct financial assistance, there exists an institutional structure for the support of new technology through which projects can graduate from government-supported research into independent commercial programmes. It is possible, particularly in the United States, but it is also perhaps true for Japan, that the political will to support marine technology was primarily based on the immediate political rhetoric of resource independence rather than on any long-term commitment to the future of these technologies. Indeed, the relative decline in support for marine technology in the United States in the 1980s tends to confirm that view. Nevertheless, it is evident that, whatever the immediate motivation, the 1970s saw considerable advances in marine technology in countries where government support was forthcoming, and that no comparable progress was made where the political will was absent.

A fourth common feature in many of the case studies was the international issues involved. These were particularly important in the context of the Law of the Sea negotiations at the United Nations where they exerted considerable influence on deep ocean mining proposals. The lack of agreement and the absence of an international legal framework during the late 1970s and early 1980s created uncertainty within the new marine industries and, with hindsight, may have been influential in the failure of manganese nodule mining (and other deep ocean projects) to progress further. Subsequently, there has been evidence, witness the hydrothermal sulphide deposits of the north-east Pacific, that resources available within the now widely recognized two hundred-mile exclusive economic zones are becoming increasingly appealing since the international aspect is absent. The confused state of international agreements on the Law of the Sea is unlikely to improve very quickly. During 1985 some Western countries gave notice that they might be prepared to withdraw from the voluntary moratorium on dumping certain categories of low-grade radioactive waste at sea. The agreement to refrain from dumping had run for several years and although its collapse does not necessarily signal a free-for-all for disposing of radioactive material at sea, it does not augur well for any multilateral international agreement relating to the use of the deep oceans. The withdrawal of certain countries from the moratorium would clearly indicate their preference for a deregulated approach on all matters of the Law of the Sea. In view of the adverse effect of uncertainty in the last few years on marine technology, it is not clear that deregulation will benefit its long-term development since uncertainty may only be perpetuated. The argument for deregulation is that the removal of restrictions will encourage early participation in new marine technology and therefore confer commercial advantages upon those firms which were in first. However, as we have seen in the case studies, the decision to participate in a new technology and the timing of that entry is a very complex matter,[1] and the influence of single factors, such as deregulation, should not be over-emphasized.

The fifth recurrent feature in the case studies has been size. Political support and public visibility seem to have been proportional to the scale of the proposal. The large, exotic and expensive seem to hold the centre of the stage and to some extent this distorts the image of what constitutes marine technology. Smaller and less ambitious projects, such as the mining of Red Sea muds or dredging for phosphorite nodules, have achieved notable technical successes and have demonstrated encouraging commercial possibilities, but have not generated the public awareness that has followed the huge schemes for floating airports or for harnessing the ocean currents. In practice, of course, the large schemes have shown less likelihood of commercial development than the smaller ones. Indeed, the contingencies of bringing large-scale projects into operation have often required the acknowledgement that a reduction in size was appropriate. OTEC, for example, had been conceived as a large-scale programme, but its best chance now of commercial deployment within the century depends upon the operation, not of a 400 MWe floating installation, but of smaller, land-based facilities. The general rule appears to have been that relatively small and relatively simple technologies aimed at specific sites and specific needs are more likely to have an impact in the near future. It is a grave error to confuse the scale of the proposal with its potential for success.

Sixthly, most proposals have been presented as self-contained operations. Yet in several cases it has been evident that there was a scope for an integrated installation including two, or even more, separate operations. The benefits of integration may be considerable in improving the overall economic performance of the schemes included in it. For example, pumping large volumes of water is expensive and, if at all possible, the duplication of costly pumping should be avoided. The water pumped by an OTEC facility might therefore be utilized by another process which would otherwise need to do its own expensive pumping. Extracting the energy from seawater does not preclude the subsequent extraction of nutrients (for mariculture) or dissolved minerals from the same body of water. The cost of pumping in such cases, would be shared between two or even three operations and so would reduce the average cost of pumping to each operation separately. Little was made of this opportunity in detail and proposals tended to be assessed in isolation, but the prospects for several new marine technologies might be significantly enhanced if the opportunities for integration were explored thoroughly. Some combinations may be mutually exclusive, but others would, at least in principle, permit collaborative developments.

The final feature that has to some degree been characteristic of all these projects is, it has to be admitted, a basic lack of commercial progress. Impressive technical progress in R & D phases has rarely been followed by comparable advances beyond demonstration and pilot plant development. This can be understood in relation to the interaction of the other common features. The continuing absence of a clear commercial future, the uncertainty over the international dimension of deep ocean development and the volatility of political support have all combined to inhibit firms from committing resources to full-scale deployment. Given these factors, which have prevailed for the last decade or so, firms are adopting a necessarily cautious approach. Another problem, which has undoubtedly contributed to the apparent lack of progress, is that the initial programmes were invariably proposed on a very large and ambitious scale. Failure in comparison with those early schemes should not be too surprising, even if the more encouraging climate of the late 1960s and early 1970s had been maintained it would have been a considerable technical and commercial achievement to have realized those early, ambitious aims. However, even for the more modest levels to which many programmes realigned themselves in the early 1980s, success has not been as evident as might have been anticipated.

The seven features discussed above, as well as the ever-present marine environment, define the context within which new marine technology must operate and within which we must assess its future. The mixture of ingredients differs beween individual projects and there has been no simple model which can be applied to them all. It is not, of course, this complex range of factors which poses the problem when assessing the future for marine technologies. All new technology is beset by a complex interaction of the political, legal and environmental with the technical and commercial. Marine technology is not unique in facing this, but with a new marine technology there is that much more uncertainty because there is no past experience upon which to call. There have not been many projects comparable with OTEC or deep sea mining or floating airports and the experience of other marine operations, such as offshore oil and gas exploration, while useful, is limited in the help it gives in dealing with the problems of these very ambitious undertakings. In a sense the context of new marine technology has yet to be defined and each project faces problems for which no precedent exists. This is, perhaps, the real challenge of new marine technology that so many of the problems, even trivial ones, have to be solved from first principles because there is no comprehensive body of past experience on which to draw. In fact it is for the technical questions that there is the most complete background of knowledge against which to work and it is surely no coincidence that the technical aspects of the programmes have often been the most successful, whereas the legal status of an artificial island or the ownership of minerals on the sea-bed or the regulation of marine disturbance from the dredging or the assessment of how much a premium the market will stand for the supply of strategically secure raw materials and energy are all the kind of question for which so little previous experience is available to aid in making decisions. But the marine technologies are raising many questions of this type. Faced with these circumstances, governments and companies, not surprisingly, believe a cautious approach is prudent until some of these issues begin to be clarified. Perhaps when this has begun it will be possible for decisions to be taken more confidently and for progress to be made more rapidly. What is perfectly clear is that the resolution of these questions from outside the technical realm is equally important to the future of marine technology as are results from R & D programmes.

If we are determined to make a general characterization of the status of the new marine technologies in the mid-1980s, it will be as a group of schemes which have demonstrated varying degrees of technical viability, but which are all now proceeding slowly as they await developments by which to assess the extent of their commercial prospects. Until the future looks a little less uncertain the commitment of further resources will be withheld. To the question of that commercial future and the prospects for the marine technologies over the new two decades we now turn.

PROSPECTS

By the expression 'future prospects for marine technology' it should be evident, from all that has gone before, that we mean the likelihood of a new technology achieving significant commercial deployment. As the Summary concluded, from the position in the mid-1980s, few of the new marine projects seem likely to achieve any widespread deployment before the end of the century. Yet this is in strong contrast to the predictions that had been made in the mid-1970s, when the opportunities for the technologies appeared almost limitless. In order to be able to offer a balanced assessment of the prospects in the next two decades it is important to account for the wild fluctuations of marine technology's fortunes in the past two decades. Only then might it be possible to see if there are reasons for being more optimistic about the future prospects of marine technology than that which the current situation implies.

There is no doubting the inventiveness and imagination which generated the proposals for marine technologies. There can be no doubting either that the late 1960s and early 1970s saw a large commitment of resources to those programmes and to a period of high expectations. But since the late 1970s interest in marine technology has grown less rapidly than in the preceding decade. The end-result of this relative decline is reflected in our conclusion that for all but technically unsophisticated and small-scale projects the future is not encouraging. It is possible that the enthusiasm in the 1970s was illusory. In America, perhaps, research programmes may merely have been used to provide access to government finance. Certainly in the case of OTEC and other large-scale marine energy programmes the rise in interest from corporations like TSR and Lockheed, both of which were traditionally involved in high technology aerospace research, seemed to correspond to a period of decline in defence contracts under President Carter. But if that was one element, it was nowhere near the whole story. Throughout the late 1960s concern was growing about the depletion of non-renewable energy and mineral resources; this provided a stimulus to the search for alternative and, if possible, renewable resources. The oil price rises of the early 1970s served to reinforce this feeling, particularly in the West, that new and secure supplies of energy and raw materials were an absolute necessity. But since then depressed growth in the Western economies has produced a glut of oil and has done much to reduce the political threat from the disruption of oil supplies. The resource extrapolations in general, were based on the growth rates that applied to the two decades of reconstruction following the Second World War and such rapid industrial growth has not been present in the subsequent two decades. Resource depletion therefore, and the political manipulation of resource supplies have become of far less concern and conventional resources seem better able to keep pace with requirements. Not surprisingly, the impetus for exotic and expensive marine alternatives has been reduced. On land, we should remember, even nuclear power, which after thirty years of operation is hardly a new technology, has not been able to make an unequivocal case for itself as an essential part of energy supply in order to counter the anti-nuclear lobby. So it is even more difficult for technologies not yet in operation to claim that they are crucial to the future supply of energy. As the threat of economic collapse caused by limitations in energy and material supply has receded in the past decade, so too has the single most powerful argument in favour of the development of alternative resources of marine origin.

Proponents of new marine programmes obviously claim that this change in attitude is shortsightedly complacent. The situation could change again and alternatives should therefore be pursued, even in the absence of an immediate need. But the case studies clearly demonstrated that, whether or not marine technology's supporters prove to be justified in their warnings, single-factor arguments will not change the future of marine technology. At present, in most countries and for most raw materials and energy supplies, the extent of political concern over secure supplies is insufficient to overcome those factors which militate against the commitment of resources to expensive alternatives.

The central problem in assessing the future of marine technology is the distortion caused by the special circumstances of the late 1960s and early 1970s. The events of the last few years have shown that marine technology as a whole will not achieve the levels of progress predicted during that period of early optimism. This does not mean that there is no future for marine technology, but its immediate future needs to be seen in the context of the mid-1980s and not the early 1970s. We have already seen that there is a form of marine technology that is appropriate to the current decade. Earlier ambitious schemes are no longer tenable, but smaller-scale developments, at the right site, may have a future. If we extrapolate into the early years of the next century on the basis of the situation in the 1980s, although it is only a short base from which to assess future prospects, it probably offers a more reliable forecast than that available from continued reliance upon the older mode of marine technology project. Circumstances could change again and conditions similar to those of the early 1970s could prevail once more, but it is unlikely. Supply problems may be experienced periodically for individual raw materials and the possibility of a new oil crisis can never be ignored, but there is not the background of high growth rates in the Western economies that characterized the decade prior to that first phase of marine technology development. Unless there is a return to long-term high growth rates in the Western economies based on resource- and energy-intensive industries it is difficult to see how resource restrictions, whether due to market forces or to political intervention, will have the same impact in the future as they did in the late 1960s and early 1970s. That is not to say that the search for new reserves will not continue, but the specific circumstances which gave impetus to marine technology as a significant option for the medium term may not be repeated.

Even if, for whatever reasons, conditions were to change rapidly such that they resembled those of the late 1960s and early 1970s, the inertia of decisions and commitments made during the past five years would restrict the rate of response; there could be no sudden return to the scale of the programme of the 1970s. Even if political and economic circumstances changed dramatically, the immediate impact on the prospects for marine technology may not be so great as might be expected. So, in seeking to assess the overall development opportunities for marine technologies in the next two decades, we must be influenced by the recent past rather more than by the aspirations of programmes proposed under the very different conditions of fifteen or twenty years ago.

The experience of the 1980s suggests that the marine technologies that do achieve commercialization in the next two decades will do so where a combination of factors renders the operation able to produce the energy or the mineral at a cost comparable with other sources for a specific market. The current resource climate implies that this criterion will not generally be met and so the extent of market penetration by resources of marine origin will be restricted. But from the successful operation of some smaller schemes, experience can be gained and the possibility of wider application will improve. Although the scenario of rapid large-scale expansion has been rejected it does not mean that there is no long-term future for the larger projects. But that future will depend on the success of these early attempts and the experience gained thereby. The development of the larger schemes, if it does take place, will now be slower than its enthusiastic supporters were predicting in the early 1970s, but growth will be steadier and will probably have better prospects than if it had been the volatile boom industry of the 1990s. By the early years of next century the mining of Red Sea muds, the dredging of phosphorite nodules, the construction of artificial islands and the operation of small, land-based OTEC installations may all be pursued commercially and if they are successful the prospects for the other schemes will be considerably improved.

The one country that might be prepared to advance its development programmes rather more quickly is Japan. We have seen that in many marine technology schemes Japan has been involved in serious research for more than a decade and although a later entrant than America it is now at the forefront in several areas. However, when assessing Japanese progress it is important, though not easy, to avoid preconceptions about Japanese technological prowess. The economic and technical history of the second half of the twentieth century will record the spectacular industrial expansion of Japan. The penetration of Western domestic markets by Japanese products testifies to good product development and aggressive marketing, but that market success is also commonly seen as the reward for commitment to R & D and for a willingness to take risks in high technology. By contrast it is assumed that the absence of such commitment to R & D largely accounts for Western, and in particular British, failure to compete with Japan in world markets. No doubt there is considerable truth in this interpretation, but the prevailing image of Japanese industrial dynamism can obscure important elements in the assessment of Japan's current position in marine technology development.

By analogy with Japanese performance in other high technology enterprises it is easy to perceive that country's commitment to novel marine technology as a further example of its positive approach to R & D and to assume, therefore, that it will reap rich rewards in the next century because of these projects. Certainly Japan is deeply involved in marine technology, but its motives for this appear to be predominantly domestic. Japan needs guaranteed supplies of raw materials and energy if its industrial growth is to be sustained. Among the major industrial nations Japan is uniquely lacking in natural resources. This is why it must explore all alternative supplies, including marine ones.

Throughout this book, the case studies have demonstrated that technical competence is insufficient to ensure the commercial development of marine technology. If a process is to compete with conventional production it must be able to deliver its product at a comparable price. In most cases that economic criterion cannot be met. Japan, however, is prepared to contemplate paying above the general market prices to secure supplies. Political will and commercial pressure combine to make Japan finance expensive research programmes, which include marine technologies, as part of a general search for alternative resource supplies. These pressures are not being felt to the same degree by the Western nations. The mere fact that Japan experiences these pressures does not mean that other countries have to follow suit, however strong the temptation may be. If the temptation is strong it is in no small way due to the image we now carry of Japanese technological success, but to follow merely because that is the course taken by Japan would be to let the image dominate over more rational analysis.

Indeed, rather than rushing in across a whole range of new technologies, the appropriate strategy for Western companies may well be to delay entry. The timing of entry into fields of new technology is a crucial decision in the management of innovation. By waiting before making the commitment to enter a new field, it becomes possible to enter with a second or third generation of technical development and to leap-frog ahead of those who have been in the field longer but who may be committed to an earlier generation of technology. Clearly there are serious commercial penalties for delaying too long, but the experience of innovation in other technologies has shown that too early is as bad as too late. The best strategy for Western companies may be to wait, to observe developments and to enter as individual technologies become viable. The experiences gained in this piecemeal approach will be valuable in supporting subsequent entries into more exotic technologies. The question of timing is vital and Western firms must make that decision for themselves, but they should be aware that too much emphasis on Japan's progress could lead Western companies to make a premature commitment. It is important that Western firms appreciate the context of the Japanese R & D effort. If they seek to emulate the Japanese commitment it must be for reasons appropriate to their own situations and not just to keep up with the Japanese.

A watching brief may be a good policy for those firms in Britain and North-Western Europe that have gained experience in the North Sea oil and gas industry. If they choose their field and their entry to suit that experience they are well-equipped to participate in an expansion of marine technology. The opportunities for these companies will be partly for large construction and heavy engineering projects. The pioneer work under-taken in Britain, Holland and Norway on the construction of concrete platforms for the offshore oil industry in the North Sea will serve as important experience if artificial islands are built in the North Sea. But opportunities for large-scale fabrication projects are limited. Economically it is sensible to undertake heavy construction work near to the site of intended operation and the locations of many new marine technologies will be remote from North-Western Europe. For example, a concrete OTEC platform could almost certainly be built in Britain by one of the consortia that built concrete oil production platforms, but it is unlikely to be worth-while since OTEC installation would need to be towed several thousand kilometres into tropical waters for deployment. It is therefore not justified to place too much dependence on the potential of large construction projects as the main opportunities for participation in marine technology. The better prospects may be found in the supply of components and in the service aspects of offshore operations, particularly maintenance and logistical support for installations: this is a large and valuable sector of marine industry. British firms have performed well in the services supplied to North Sea oil and gas development and have gained experience in the operation and peformance of structures and materials in a harsh marine environment over a period of some twenty years. The techniques developed there will have applications farther afield and the supply of specialized components and services is not confined to the novel marine technologies. Developments initially undertaken for new marine projects may find a second market in other marine operations, such as fisheries, shipping and, of course, in other offshore oil and gas operations beyond the North Sea. The potential market is therefore considerably larger than one confined solely to the novel technologies. Developments in navigation, station-keeping, safety, maintenance, personnel transfer, etc. can all be applied to other marine activities.

For firms in both Britain and elsewhere in North-Western Europe which have gained experience in the North Sea, new marine technologies will offer scope for expansion and new market development. This market may not begin to expand until the last few years of the century and individual firms must make their own commercial decisions concerning their entry into the field. But it is important that their decisions should not be falsely guided either by the residue of over-optimistic predictions from America in the 1970s or by the current level of Japanese research activity. Neither of these accurately reflects the present prospects for marine technology development. This is not to ignore the possibility of a government, perhaps in Western Europe, choosing to pursue marine resource exploitation as a high priority and to reinforce that decision with funding for R & D. In these circumstances firms may be persuaded to enter earlier than they otherwise might. But even in this situation a firm would be wise to ensure that the government had a long-term commitment, for if it has not the premature withdrawal of government finance may leave the firm with a commitment to marine technology that the commercial climate cannot justify. This problem has been faced by some American firms since the late 1970s and disengagement from the programmes after having committed resources and having begun to gain expertise is far from easy. Indeed, there is the suspicion that some current programmes in America may be based more upon the desire to utilize the experience gained from previous government-financed projects than upon a genuine belief in the commercial future of the schemes.

It may be strange, if not actually perverse, that a book concerned with the future of new technology should expend such efforts to counsel caution and to warn against the influence of enthusiastic advocacy in the recent past. We are, it is often said, in a time of rapid technical change and the aggressive pursuit of technical change is the key to economic success for both countries and companies. But we have shown that such an approach is inappropriate to the current progress of marine technology. There is no doubt that the prospects across a broad range of projects are far less rosy than they were a decade ago. But it would be equally untrue to conclude that because of this relative decline from the optimism and intense activity of the 1970s there is no longer any future for marine technology. In this book we have sought to reassess this future in the light of the changed political and economic climate of the 1980s. In particular the impact of lower growth rates in the West, the changed attitude in the United States towards federal finance and the confused state of the negotiations concerning the legal regime relating to the resources of the deep oceans have all contributed to a reduction in the stimulus for the development of new marine industries. Nevertheless, we have tried to demonstrate that this results in a change of emphasis for marine developments rather than in their total demise.

We now suggest that the future of marine resource technology is one of slower growth, based on incremental steps, rather than on the rapid development and deployment of very large installations. Indeed, it is not improbable that the long-term prospects for a stable and broadly-based marine industry are better served by the rejection of the more exotic and extravagant plans, the failure of one of which may have done serious long-term damage to the future of all marine technologies. The danger of that now appears to be averted.

If these technologies are to be viable and to become significant contributors to the energy and mineral supplies of the twenty-first century they must now develop within the more restrained political and economic climate currently prevailing. Whether this is possible or not remains to be seen, but at least, by acknowledging that circumstances have changed fundamentally since the mid-1970s, both firms and governments should be better placed to reassess their future involvement in marine technology.

In Europe there is already some evidence of the emergence of this new attitude towards marine technology. During 1986 the European Parliament, principally through its Committee on Energy, Research and Technology, compiled a report on the establishment of an EEC marine research institute. The significance of this development is that the report has been written explicitly in the context of Europe in the 1980s and therefore it will not necessarily adopt the attitudes which influence current American and Japanese policies. Accordingly this EEC initiative represents an opportunity for a genuinely European approach to new marine technologies in the circumstances that are now prevailing.

NOTES

1. The question of timing an entry into new technology will be discussed again in more detail below. See pp. 127–9.

Bibliography

The review presented in this book is based principally upon the studies carried out by the Marine Resources Project at the University of Manchester. Full bibliographic information and details of references are included in the reports and papers that have been produced by the members of the Project, most of which have been published. A few, however, have been subject to restrictions of access.

The following select bibliography provides an introduction to the literature that is currently available from the work of the Marine Resources Project and other researchers.

SELECT BIBLIOGRAPHY

Bingham, A., 'Artificial Island Taps Wave Power', *Offshore Engineering*, 20 (8), 1979.

Borgesse, E. and Ginsberg, N. (eds), *Ocean Yearbook*, University of Chicago Press, Chicago, 1978.

Brin, A., *Energy and the Oceans*, Westbury House, Guildford, 1981.

Cameron, H., Ford., G., Garner, A., Gibbons, M. and Majoram, T., *Manganese Nodule Mining: Issues and Perspectives*, Marine Resources Project, PREST, University of Manchester, 1980.

Cameron, H., Georghiou, L., Perry, J. and Wiley, P., 'The Economic Feasibility of Deep Sea Mining', *Engineering Costs and Production Economics*, 5, 1981, pp. 279–87.

Cameron, H. and Georghiou, L., 'Production Limits—Who Benefits?', *Marine Policy*, 5 (3), July 1981, pp. 267–70.

Cameron, H., Georghiou, L., and Ford, G., 'Retrospective: The Manganese Nodule Bonanza', *Futures*, 14, December 1982, pp. 554–5.

Cameron, H., Spagni, D., Ford, G. and Taylor, E., *Artificial Islands*, Marine Resources Project, PREST, University of Manchester, 1983.

Cameron, H., Vernon, F. and Georghiou, L., 'Prospects for the Extraction of Uranium from Seawater', *Institute of Chemical Engineers Symposium*, Series no. 78, 1982.

Cannon, J. and Herman, S., *Energy Futures*, Ballinger, Cambridge, Mass., 1977.

Carliss, J. *et al*., 'Submarine Thermal Springs on the Galapagos Rift', *Science*, 203, 1979, pp. 1073–83.

Claude, G., 'Power from the Tropical Oceans', *Mechanical Engineering*, 52, December 1930, pp. 1039–44.

Crawford, D., 'The State of the Art in Exotic Platforms', *Offshore*, 39 (4), 1979, pp. 52–65.

Cronan, D., *Underwater Minerals*, Academic Press, London, 1980.

Crowson, P., *Mineral Handbook*, Macmillan, London, 1982.

Earney, F., *Petroleum and Hard Minerals from the Sea*, Edward Arnold, London, 1980.

Ford, G., 'Indian Marine Technology Policy', *Marine Policy*, 7, April 1983, pp. 122–3.

Ford, G., Cameron, H., Georghiou, L., Spagni, D. and Taylor, E., *Japan and New Marine Technology*, Marine Resources Project, PREST, University of Manchester, 1983.

Ford, G. and Georghiou, L., 'Marine Technologies for the 1990s', *Phys. Technol.*, 13, 1982, pp. 11–17.

Ford, G., and Georghiou, L., 'Japan Stakes Its Industrial Future in the Sea', *New Scientist*, 94, 1982, p. 649.

Ford, G., Georghiou, L., and Cameron, H., 'Using the Seas During the 21st Century', *Impact of Science on Society*, 1983/4, pp. 491–501.

Ford, G. and Little, D., 'Garden Under the Sea', *Far Eastern Economic Review*, 14/6/1984, pp. 118–20.

Ford, G., Niblett, C. and Walker, L., *Ocean Thermal Energy: Prospects and Opportunities*, Marine Resources Project, PREST, University of Manchester, 1981.

Ford, G., Niblett, C. and Walker, L., 'Ocean Thermal Energy Conversion', *I.E.E. Proceedings*, 130 (Part A, No. 2), March 1983, pp. 93–100.

Georghiou, L. and Ford, G., 'Arab Silver from the Red Sea Mud', *New Scientist*, 89, 1981, pp. 470–2.

Georghiou, L., Ford, G., Gibbons, M. and Jones, G., 'Japanese New Technology: Creating Future Marine Industries', *Marine Policy*, 7, October 1983, pp. 239–53.

Glasby, 'Minerals From the Sea', *Endeavour*, 3 (2), 1979, pp. 82–6.

de Groot, S., 'An Assessment of the Potential Environmental Impact of Large Scale Sand Dredging for the Building of Artificial Islands in the North Sea', *Ocean Management*, 5, 1979, pp. 211–32.

Gummett, P., Georghiou, L., Spagni, D. and Ford, G., 'When Waste Disposal is a Drop in the Ocean', *Guardian*, 17/2/83, p. 19.

Institute of Geological Sciences, *World Mineral Statistics 1977-81: Production, Exports and Imports*, NERC, HMSO, London, 1983.

Isaacs, J. and Schmitt, W., 'Ocean Energy: Forms and Prospects', *Science*, 207 (4428), 18/1/80, pp. 265–73.

Jellinek, H. and Masuda, H., 'Osmo-Power: Theory and Performance of an Osmo-Power Pilot Plant', *Ocean Engineering*, 8 (2), 1981, pp. 103–28.

Kent, P., *Minerals in the Marine Environment*, Edward Arnold, London, 1980.

Ketchum, B. *et al.*, *Ocean Dumping of Industrial Waste*, Plenum Press, New York, 1981.

Lacey, R., 'Energy By Reverse Electrolysis', *Ocean Engineering*, 7 (1), 1980, pp. 1–47.

Lissaman, P., 'The Coriolis Program', *Oceanus*, 22 (4), Winter 1979/80, pp. 23–8.

Majoram, T., 'Manganese Nodules and Marine Technology', *Resources Policy*, 7 (1), 1981.

McCormick, M., 'Wind/Wave Power Available to a Wave Energy Converter Array', *Ocean Engineering*, 5 (2), pp. 67–74.

McRobie, J. and Green, D., *Potash Extraction from The Dead Sea: A Summary* Marine Resources Project, PREST, University of Manchester, 1980.

Mochizuki, H. and Mitsuhashi, W., 'An Energy Conversion Method in the Ocean Using the Density Differences of Water', *Ocean Engineering*, 8 (1), 1981, pp. 91–6.

Mottl, M., 'Submarine Hydrothermal Deposits', *Oceanus*, 23 (2), 1980, pp. 18–27.

Olssan, M., Wick, G. and Isaacs, J., 'Salinity Gradient Power: Utilising Vapour Pressure Differences', *Science*, 206 (4417), 1979, pp. 452–4.

Ross, D., *Opportunities and Uses of the Ocean*, Springer-Verlag, New York, 1978.

Ryther, J., 'Fuels From Marine Biomass', *Oceanus*, 22 (4), 1979/80, pp. 50–8.

Salmon, R. and Harding, S., 'Gas Concentration Cells for the Conversion of Ocean Wave Energy', *Ocean Engineering*, 6 (3), 1979, pp. 317–27.

Sardar, Z., 'Red Sea States Unite Against Pollution', *New Scientist*, 89, 19/2/81, p. 472.

Simnett, J., *The Red Sea Metalliferous Muds: A New Source of Metals?* M.Sc. Thesis, University of Manchester, 1982.

Simnett, J., Ford, G., and Reeve, N., *Remote Sensing in the Marine Environment*, Marine Resources Project, PREST, University of Manchester, 1984.

Smith, F. and Charlier, R., 'Waves of Energy', *Sea Frontiers*, 27 (3), pp. 138–49.

Spagni, D. and Ford, G., 'Treasures of the East', *Far Eastern Economic Review*, 29/8/84, pp. 71–2.

Spagni, D., Ford., G. and Simnett, J., 'Sulphides: The Next Deep Sea Klondyke', *New Scientist*, 14/4/83, p. 72.

Spagni, D., Georghiou, L. and Ford, G., 'Deep Seabed Mining: The Last Chance', *New Scientist*, 97, 27/1/83, p. 255.

Taylor, E., Cameron, H. and Spagni, D., *Phosphorite Nodules: Fertilizer from the Sea*, Marine Resources Project, PREST, University of Manchester.

Toms, A., Ford, G. and Georghiou, L., *Ocean Thermal Energy Conversion: Costing A Land Based Plant*, Marine Resources Project, PREST, University of Manchester, 1983.

'Wave Power Reviewed', *Offshore Engineer*, January 1981, pp. 25–36.

There are several journals which carry papers concerned with the development of new marine technology. They include: *Marine Policy*, *Oceanus*, *Ocean Engineering*, *Ocean Industry*, *Ocean Management*, *Offshore Engineering* and *Sea Technology*. This is not an exclusive list, but monitoring these and other similar publications will indicate the progress and direction of future developments, as well as providing further information on those programmes reviewed in this book.

Index

airports 93, 103–4, 123, 125
aluminium 50, 63, 70, 71
America, *see* United States of America
Arctic 101–3, 105
artificial islands 3, 47, 80, 88, 128
    airports 103–4
    Arctic oil exploration 101–3, 105–6
    barge-mounted plants 97–9
    design and construction 94–7
    future projects 103–6
    Japan 99–101, 103, 105, 106
    United States of America 104–5
    uses 93–4
Australia 56

Baja California 29
biomass energy 87–8
Brazil 98–9

calcium 53
Canada 56
Central America 29
Chatham Rise 29, 30, 31
Clarion-Clipperton Zone 8, 20
coal, undersea 2, 89–90
cobalt 2, 7, 18
continental shelf minerals 2, 25
copper 2, 7, 18, 32, 33, 36, 38, 50, 52

Dam Atoll 83, 84, 91, 120
Dead Sea 46, 119
    minerals 46, 53–4, 57
density gradients energy 3, 79, 86, 119

East Pacific Rise 49, 52
echo-sounding 110–11
ERS Project 116
European Economic Community (EEC) 132
    European Parliament 132

ferromanganese nodules, *see* manganese nodules
ferromanganese oxides 2
fishing 2, 3, 109, 130
France 10, 72, 73

Galapagos Rise 49
gold 50, 54
gravimetric surveying 111

Great Britain 4, 10, 128, 130
    ocean thermal energy conversion (OTEC) 73–4
Group of 77 21–2
Gulf of Mexico 65, 66, 75, 76

Hawaii 20, 65, 66, 83
Holland 72, 93, 130
hydrogen 63
    production at sea 88
hydrothermal muds *see* Red Sea metalliferous muds
hydrothermal sulphides 45, 49–52, 57, 119, 122
    formation 49–50
    Japan 52
    mining consortia 51–2
    recovery 50–1
    United States of America 51–2

India 72
Indian Ocean 29
ion exchange 54–7, 119
iron 2, 7, 36, 50

Japan 5, 58, 72, 98, 99, 121, 122, 128–30, 131, 132
    artificial islands 99–101, 103, 105, 106
    biomass energy 87
    deep ocean mining 10–11
    hydrothermal sulphides 52
    marine observation satellites 116
    ocean current energy 85
    ocean thermal energy conversion (OTEC) 72–3
    uranium from sea water 56
    wave power 84

Kobe Port Island 99–101, 104

Landsat 116
Law of the Sea 4, 120, 122
    United Nations Conference on the Law of the Sea (UNCLOS III) 20–3, 107
lead 2, 36

magnesium 53
magnetic surveying 111
manganese 2, 7, 18, 50

manganese nodules 2, 7–23, 25, 32, 41, 45, 50, 51, 52, 57, 94, 105, 120, 122
  economics 19–20
  formation 7–8
  history 7
  legal 20–2
  mining consortia 8–11
  processing 17–18
  recovery 11–17
  United States of America 8–10, 21–2
METEOSAT 114
Mexico 20, 29
molybdenum 54

NASA 115–16, 117
navigation 3, 110, 130
New Zealand 29–30
nickel 2, 7, 18
North Atlantic 85
North Sea 46, 47, 74, 89, 96, 98, 103, 104, 130
  oil and gas 4, 96, 111, 130
Norway 130
nuclear power 2, 55, 64–5, 75, 93, 104–5
nuclear waste disposal 122–3

ocean current energy 85, 123
ocean thermal energy conversion (OTEC) 3, 55, 56, 57, 58, 79, 82, 83, 86, 87, 88, 91, 94, 104–5, 120, 123, 125, 126, 130
  artificial islands 104–5
  cold water pipe (CWP) 68–9
  cost estimates 74–5
  environmental impact 75–6
  France 73
  Great Britain 73–4
  heat exchangers 69–72
  Japan 72–3
  plant design 67–72
  principles 61–4
  United States of America 64–72
Ocean Thermal Energy Conversion Act 65
Ocean Thermal Energy Conversion Demonstration Fund 65
Ocean Thermal Energy Conversion (Research and Development) Act 65
ocean wind energy 88–9
oil and gas 2, 3, 93, 94, 95, 97, 101–3, 109, 125, 130
  North sea 4, 96, 111, 130
oozes, calcareous and siliceous 2
oozes, radiolarian 45, 48, 57, 120
OPEC 75
OTEC see ocean thermal energy conversion

Pacific Ocean 8, 45, 48, 49, 85, 122
phosphorite nodules 2, 3, 25–32, 45, 57, 94, 105, 119, 123, 128
  economics 30

environmental impact of dredging and processing 30–1
  formation 26–7
  legal 30–2
  New Zealand 29–31
  processing 28–30
  recovery 27–9
phosphorus 26, 53
polder 93, 95, 96
polymetallic nodules, see manganese nodules
polymetallic sulphides, see hydrothermal sulphides
positioning 110
post-war reconstruction 1, 126
potash 53–7

radiolarian oozes 45, 48, 57, 120
Red Sea metalliferous muds 3, 32–41, 45, 52, 119, 123, 128
  formation 32–3
  metal demand 38
  metals and minerals 32
  political context 38, 41
  processing 37–8
  recovery 33–6
  safeguards for environment 36–7
  structure of muds 33
remote sensing by satellite 112–16
resource survey techniques 110–11
  acoustic 110–11
  field 111
  satellite 112–16

salinity gradients energy 3, 79, 85–6, 119
sands and gravels 2, 27, 45, 46–7, 57, 120
satellites
  navigation 3, 17
  surveying 112–16
Saudi Arabia 32, 38–9
Saudi-Sudanese Joint Red Sea Commission 33, 39
Scripps Institute of Oceanography 48
Seasat 115–16
shipping 2, 109, 130
silver 2, 32, 33, 36, 38, 50, 51, 52
sodium 53
Sudan 32, 39
Sweden 72

tidal power 80–2
titanium 70, 71, 72

United Nations Conference on the Law of the Sea (UNCLOS III) 20–3, 107
United States of America 3, 4, 5, 8–9, 61, 91, 122, 126, 128, 131, 132
  artificial islands 104–5
  Department of Energy 66, 67

domestic legislation 21–2, 29, 31
hydrothermal sulphides 51–2
manganese nodules 8–10, 21–2
National Oceanographic and Atmospheric
    Administration 115
ocean thermal energy conversion (OTEC) 64–
    72
Office of Technology Assessment 69, 74
space programme 4, 115–17
wave power 83, 84
Upwelling 27

uranium 2, 54, 55, 56

vanadium 54

wave power 82–5
    Dam Atoll 83–4
    Japan 84
    United States of America 83–4
West Germany 10

zinc 2, 32, 33, 36, 38, 50, 51, 52